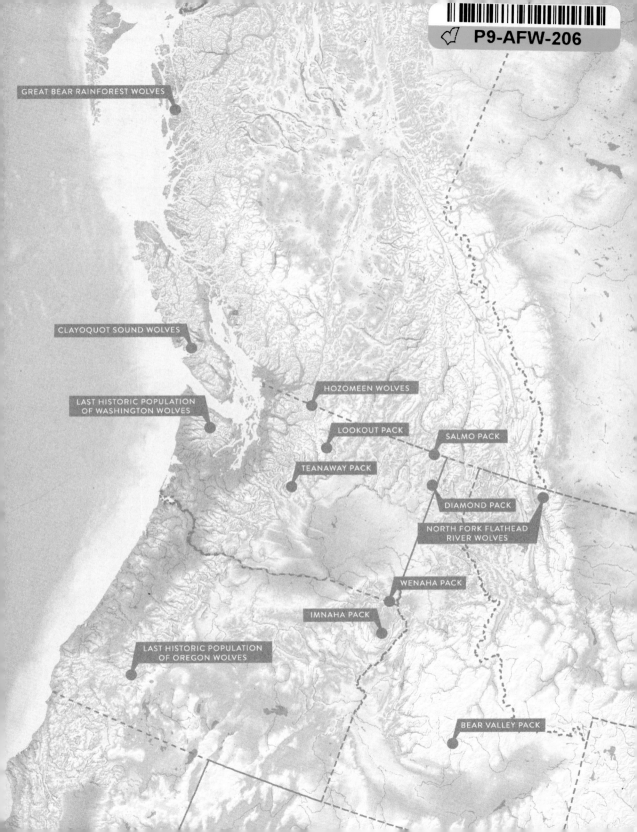

GREAT BEAR RAINFOREST WOLVES

P9-AFW-206

CLAYOQUOT SOUND WOLVES

HOZOMEEN WOLVES

LAST HISTORIC POPULATION
OF WASHINGTON WOLVES

LOOKOUT PACK

SALMO PACK

TEANAWAY PACK

DIAMOND PACK

NORTH FORK FLATHEAD
RIVER WOLVES

WENAHA PACK

IMNAHA PACK

LAST HISTORIC POPULATION
OF OREGON WOLVES

BEAR VALLEY PACK

WOLVES
IN THE LAND OF
SALMON

WOLVES
IN THE LAND OF
SALMON

DAVID MOSKOWITZ

TIMBER PRESS
PORTLAND • LONDON

FOR MY UNCLE LOUIS GOLDFARB,
FOR HIS RESTRAINED
AND GENTLE MENTORSHIP

Frontispiece: A wolf crosses a lead of water along the shore of an
island in Clayoquot Sound, British Columbia.

Copyright © 2013 by David Moskowitz. All rights reserved.

All photographs by David Moskowitz except page 33 by
Conservation Northwest and page 202 by the Oregon
Department of Fish and Wildlife.

Illustrations on pages 56–57 by Jenn Wolfe;
all other illustrations by David Moskowitz.
Maps created by Analisa Fenix / Ecotrust under a Creative
Commons license and prepared for publication by Laken Wright.

Published in 2013 by Timber Press, Inc.

The Haseltine Building
133 S.W. Second Avenue, Suite 450
Portland, Oregon 97204-3527
timberpress.com

6a Lonsdale Road
London NW6 6RD
timberpress.co.uk

Printed in China
Book design by Laken Wright
Second printing 2013

Library of Congress Cataloging-in-Publication Data

Moskowitz, David, 1976–
  Wolves in the land of salmon / David Moskowitz.
    p. cm.
  Includes bibliographical references and index.
  ISBN 978-1-60469-227-3
  1. Wolves—Northwest, Pacific. 2. Wolves—Northwest, Pacific—
Pictorial works. I. Title.
  QL737.C22M676 2013
  599.773—dc23
                        2012025205

A catalog record for this book is also available from the British
Library.

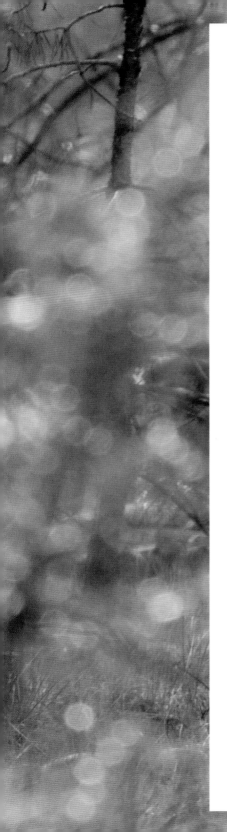

CONTENTS

A Northwest Wolf Chronology   6

INTRODUCTION: See for Yourself   10

1. NORTH CASCADES:
   Finding Their Way Back   24

2. OUR SOCIAL CARNIVORE:
   Evolution and Biology   50

3. THE SELKIRKS AND COLUMBIA
   HIGHLANDS: Where the Rainforest
   Meets the Rockies   80

4. THE OPPORTUNISTIC WOLF:
   Diet and Hunting Behavior   108

5. RAINFOREST WOLVES: British
   Columbia's Central Coast and
   Vancouver Island   134

6. WHERE DID THEY COME FROM
   AND WHERE ARE THEY GOING?   168

7. BLUE MOUNTAINS: Wolves, Elk,
   and Cows in Northeastern Oregon
   and Central Idaho   196

8. SHAPESHIFTER: The Changing
   Relationship between Humans
   and Wolves   230

9. ISOLATION: Lessons from the
   Olympic Peninsula and Beyond   254

EPILOGUE: Brave New World   288

Notes   294
Bibliography   303
Acknowledgments   324
Index   327

MAPS

Wolves in the Pacific Northwest   endpapers

The Pacific Northwest   17

North Cascades   28

Columbia Highlands, Selkirk Mountains, and Northern Rocky Mountains   84

British Columbia Coast   138

Wolf Distribution in the Pacific Northwest   175

Wolf Population Sources and Dispersal in the Pacific Northwest   179

Wolf Subspecies in the Pacific Northwest   181

Blue Mountains and Salmon River Mountains of Northeastern Oregon and Central Idaho   200

Coastal Washington and Oregon   258

Approximate Coyote Distribution in the Pacific Northwest   283

A NORTHWEST WOLF CHRONOLOGY

1.8 to 2.5 million years ago	Wolves and coyotes depart from their shared ancestor, *Canis leophagus*, in North America; precursors to modern wolves migrate to Eurasia.
800,000 to 300,000 years ago	*Canis lupus* emerges in Eurasia.
130,000 to 100,000 years ago	*Canis lupus* migrates to North America from Eurasia.
13,000 to 9000 years ago	Continental ice sheet retreats from the Pacific Northwest.
14,000 to 8000 years ago	Pleistocene extinction event occurs at the end of the last period of glaciation; North America loses about half of its large mammal species.
10,000 to 8000 years ago	Dire wolves (*Canis dirus*) becomes extinct, and *Canis lupus* emerges as the dominant large social carnivore across North America.
early 1800s	First historical regional documentation of wolves eating salmon; a die-off of wolves in the Oregon Coast Range is attributed to a parasite which wolves contract from eating salmon.
1821	Hudson's Bay Company establishes fur trading posts in the Northwest.
1843	Bounties are placed on wolves in the Oregon Territory.
1850 to 1900	Wolves become scarce across most of Washington and Oregon. Populations of elk and many other carnivores and game animals decline similarly across the region.
1909	U.S. President Theodore Roosevelt creates Olympic National Monument to protect remaining Roosevelt elk; wolves are already scarce in the Olympic Peninsula.
1910	Oregon issues a moratorium on elk hunting; wolves are close to extinct in the state.

1920	Last confirmed wolf specimen is collected in Washington State from the western side of the Olympic Mountains, though reliable sightings in the Olympics are reported in the 1920s, 1930s, and even as late as the early 1950s.
1930s	The last consistent reports of wolves in Oregon document them on the west slope of the Cascades.
1933	Following the rebound of elk populations, Oregon initiates a carefully regulated elk-hunting season.
1938	U.S. President Franklin Roosevelt expands Olympic National Monument into the current Olympic National Park.
1946	Last bounty for a wolf in Oregon is paid for an animal taken from Umpqua National Forest on the western slopes of the Cascades in southern Oregon.
1950s to 1970s	Wolf populations in the Pacific Northwest probably hit their lowest levels and most contracted range. Wolves are functionally, if not completely, extinct from Oregon, Washington, Idaho, Montana, and southern portions of British Columbia.
1960s	British Columbia's wolf eradication programs destroy—either completely or nearly so—Vancouver Island's wolf population.
1970s	Parvovirus is introduced into wild canids in North America from domestic dogs and is now endemic across the continent. Wolves begin reestablishing themselves in the southern portion of the Rockies in Canada following the abatement of control efforts in British Columbia and Alberta.
1978	Wolves are listed as an endangered species and receive federal protection across the lower forty-eight U.S. states under the Endangered Species Act.
1986	The North Fork of the Flathead River in northwestern Montana becomes the site of the first documented breeding pack to return to the northwestern United States, comprising wolves that had dispersed from over the border in the Canadian Rockies.

1990	Washington Department of Fish and Wildlife (WDFW) documents a pack of wolves with pups during a howling survey at the north end of Ross Lake, in North Cascades National Park. Though evidence of wolves here continues sporadically, further documentation of breeding wolves in the area is lacking for the next two decades.
1995 and 1996	U.S. Fish and Wildlife Service (USFWS) translocates wolves from the Canadian Rockies in Alberta and British Columbia to central Idaho and Yellowstone National Park.
1999	The first wolf to disperse to northeastern Oregon from Idaho is documented, a female who is captured and returned to Idaho.
2000 to 2007	Three more wolves from Idaho are found dead in northeastern Oregon, two of them shot.
2006	B300, a young female, is trapped and radio collared in western Idaho.
2008	WDFW traps and radio collars the breeding male and female of what became Washington's first documented pack, the Lookout pack, on the east slope of the North Cascades in north-central Washington.
2009	Wolves are removed from the federal list of endangered species in the northern Rockies for the first time. Federal status as endangered continues in California and the western two-thirds of Washington and Oregon.

B300 and her mate, also from Idaho, establish Oregon's first modern wolf pack—the Imnaha pack—in the Wallowa Mountains in northeastern Oregon.

The Diamond pack is discovered as Washington State's second confirmed pack. |
| 2010–2011 | The Lookout pack's alpha female disappears in spring. By winter of 2010-2011, only two animals, the alpha male and a second smaller animal, likely one of his offspring, are documented in the pack's territory. |
| 2011–2012 | Washington and Oregon estimate wolf populations of about thirty animals in each state. Idaho initiates an aggressive hunting and trapping season, reducing the state's wolf population by about 50 percent. |

INTRODUCTION

See for Yourself

HEN I WAS TWELVE, about a week before my school science project was due, my mother discovered that I had decided to forgo carrying out the experiment which was the basis of my project. Instead I fabricated the data and results. The project was about probability. The experiment required the use of a quincunx, a device that involves dropping marbles down a uniform pattern of pegs starting at the apex of a two-dimensional pyramid. Theoretically, most of the marbles should land toward the center of the base of the quincunx, with progressively fewer toward the edges, forming a bell curve, a common pattern of mathematical distribution in many natural systems.

Being twelve, I figured that since many folks had done this before, why should I go through the trouble of repeating the experiment? I could take their design and results. I wrote up all the components of the virtual experiment and made graphs that showed my marbles landing in an exact bell curve with the greatest number in the center and progressively fewer as you moved away from it on either side evenly until barely any landed on the edges. Unaware that anyone might object to this, I mentioned it when my mother asked about the project a week before it was due. My mother, a medical doctor and university professor actively engaged in a variety of research projects, was absolutely horrified that her son was on the verge of completely fabricating the results of the first scientific endeavor of his life. Thankfully, she took the opportunity to introduce me to a world of inquiry that has greatly influenced my passions and curiosities ever since.

A week later, after the actual construction of a quincunx and many boring hours of dropping hundreds of marbles into it and recording the results (aided by my eternally patient grandmother), my initiation into scientific investigation had begun. My results, which became graphs and discussion points for my poster, showed that, while the preponderance of my marbles did fall toward the center, they did not produce a perfect bell curve. Variations between one side and the other existed; some slots farther from the center had more landings than others closer to the center. My results supported the general pattern while also demonstrating that reality seldom matches theory exactly. My project went on to be selected for the county science project contest, and I learned a valuable lesson about how the world works.

Since this early experience, as both a naturalist and an engaged citizen in a democracy, I've always considered it prudent to educate myself and critically analyze both the natural and social world around me. This includes going out and seeing things for myself before making my mind up about how I believe things work and the appropriate ways to proceed. It is in this spirit that I've built this book around stories from the field and around my attempts to unravel the complex, sometimes counterintuitive, and almost always politically charged story of the relationship between the wolves, wildlands, and humans of the Pacific Northwest.

Canis lupus, the largest member of the canid family in the world, has an impressive evolutionary track record and an exceptional history of influence on human cultures across the northern hemisphere, where the two species' ranges have overlapped for millennia. Weighing in at an average of about ninety pounds for males and eighty pounds for females in the Pacific Northwest, adult wolves stand waist high to the average person, a substantial physical presence. Their intelligence, highly social nature, and adaptable behavior have further contributed to our fascination with the animals whose howls echo with haunting beauty through the wild landscapes they inhabit and through the myths and stories of human cultures around the globe.

Amazingly, despite their relatively large amounts of press here in recent years, wolves are living in the region in quite limited numbers. In Idaho fewer than nine hundred animals were reported in 2009, their peak census in that state. To keep the species from being relisted as endangered, Idaho has to maintain a minimum of only about one hundred wolves. As of the summer of 2011, the Washington Department of Fish and Wildlife (WDFW) estimated Washington's wolf population at about thirty; Oregon estimated a similar number as of December 2011. As the benchmark for removing wolves from their state endangered species lists, both Oregon and Washington use breeding pairs, a term legally defined as a male and female wolf and at least two offspring which survive to the end of a given year. Translating this into an estimated total number of wolves, the state recovery and management plans in Oregon allow the delisting process to begin once the state has seven breeding pairs, or an estimated census of as few as about seventy wolves in the state. In Washington fifteen breeding pairs, about 210 wolves, would be required for complete delisting from the state endangered species list. Meanwhile the actual numbers of wolves will fluctuate widely at the whims of hunting and management plans for states and provinces in the region. In Idaho's 2011–2012 wolf-hunting season, hunters and trappers took more than 375 wolves, more than half the estimated population of 746 for the entire state, as the state attempted to reduce wolves to closer to the minimum numbers required to keep the animal off the federal endangered species list.

How did people, wolves, and wildlands get to this juncture? The long history of relationships among the characters in this story begins about thirteen thousand years ago with the retreat of the last period of continental glaciation. We often think of contemporary events from the perspective of human politics.

PREVIOUS: Fresh snow blankets higher elevations in the Blue Mountains in the home range of the Wenaha pack in northeastern Oregon.

But focusing instead on the ecological and evolutionary relationships between humans and other species—*Homo sapiens* and *Canis lupus*, for example—and on their relationships with their surroundings can shed some much-needed light on these groups and on the inner workings of natural systems as well.

While the return of wolves to the east in the Rockies has been in the news for decades, wolves are just now getting their moment in the spotlight in the Pacific Northwest. The story here will be a different one, though. In a geographic landscape as defined by the ocean as the Pacific Northwest is, even a terrestrial species such as the wolf can't help but be influenced. Similarly the unique human cultural landscape of the Pacific Northwest is shaping the way that residents of the region think and feel about the return of this compelling species.

The rain-soaked Pacific Coast and adjacent dense temperate rainforests at the western edge of the region may be the iconic landscape here, but the Pacific Northwest in its totality might be better defined by the geographic and ecological extremes it encompasses. At its southern boundary in northern California, Mount Lassen and Mount Shasta mark the southern tip of the Cascade Range and temperate rainforests cloak the coast. To the north along the coast, starting around central British Columbia, the region fades into the boreal ecosystems that define northern British Columbia and Alaska.

Inland the Cascades and British Columbia Coast Range create a major ecological dividing line. East of these ranges the moderating influences of the ocean are greatly reduced and a continental climate prevails. Yet the ocean continues to define the reach of the region through its emissaries: salmon and the rivers that connect them to the interior. The Columbia and Frasier rivers stretch far inland, allowing salmon and marine nutrients to travel hundreds of miles into the continent. The watersheds of these two rivers, along with the Rocky Mountains to the east and Great Basin Desert to the southeast, mark the region's interior boundaries. Reminders of the region's classic coastal ecosystems can also be found in the interior, with patches of inland temperate rainforest dotting the eastern edges of the Columbia River's watershed in British Columbia, Washington, and Idaho. In total, the Pacific Northwest includes everything from rainforests which receive over one hundred inches of rain a year, to deserts which get less than ten. With sea-level forests on the coast, just miles away alpine tundra dots the high peaks of the mountains, and glaciers linger in high-elevation cirques. Dry conifer forests, grasslands, and shrub steppe dominate interior arid landscapes.

I've approached my research and photography for this book in a manner consistent with my training as a wildlife tracker. When I moved from Ohio to northern California as a teenager, I became fascinated with the natural world. Perhaps

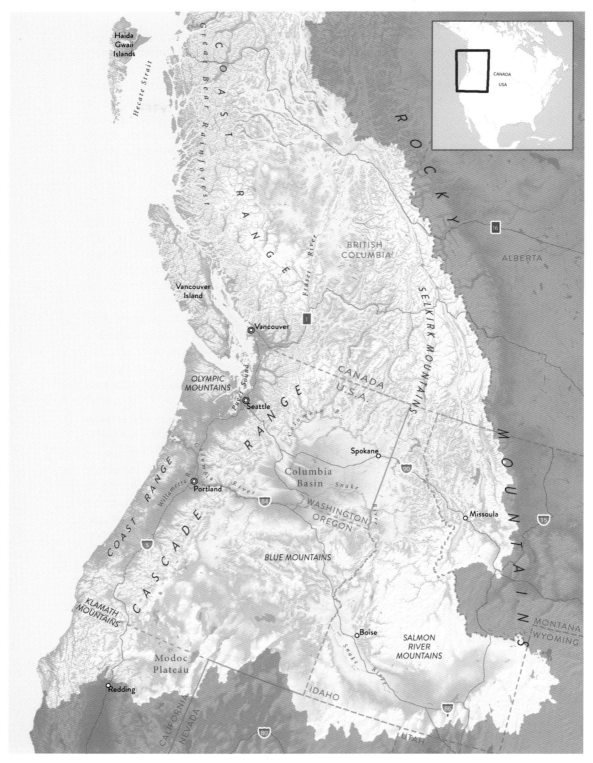

Haida
Gwaii
Islands

Hecate Strait

Great Bear Rainforest

C O A S T

Fraser River

BRITISH
COLUMBIA

ROCKY

16

ALBERTA

SELKIRK MOUNTAINS

R A N G E

Vancouver
Island

1

Vancouver

OLYMPIC
MOUNTAINS

Puget Sound

CANADA
U.S.A.

Seattle

R A N G E

Columbia R.

Spokane

90

M O U N T A I N S

COAST RANGE

Columbia River

Willamette R.

Portland

84

Columbia
Basin

Snake River

WASHINGTON
OREGON

5

Missoula

15

BLUE MOUNTAINS

KLAMATH
MOUNTAINS

C A S C A D E

Boise

SALMON
RIVER
MOUNTAINS

MONTANA
WYOMING

Modoc
Plateau

Snake River

IDAHO

86

Redding

CALIFORNIA
NEVADA

80

UTAH

CANADA
USA

THE PACIFIC NORTHWEST

inspired by my early lesson in the value of inquiry and direct observation, I was particularly drawn to the field of wildlife tracking. Tracking is a challenging pursuit in which the practitioner is constantly left with incomplete evidence and secondary signs of the research target. Trackers make detailed observations of the physical clues left on the landscape by the passage and activities of animals. To piece together the animal's story, the tracker combines observations, past experiences, and knowledge of the focal species and its relationship to the environment. Tracking requires you to integrate physical observation skills, prior knowledge, inductive and deductive reasoning, intuition, and imagination.

In his 1990 exploration of wildlife tracking, based on his fieldwork with the San bushmen of the Kalahari in southern Africa, Louis Liebenberg distinguished the various physical and mental requirements of the art of tracking. At its simplest level, such as determining the species identity of a track discovered on the edge of a stream, only simple inferences may be required. About how big was the animal that left this track? How does the shape of the imprints of toes and palm compare with the form of feet of that size belonging to various creatures that might be found in such an area? On a broader level, however, as one attempts to uncover the specific behavior of an animal, follow its trail through the forest, or understand how it responds to and influences its environment —the sorts of questions we are asking in this book—the tracker must blend his or her physical observations and knowledge to develop a plausible story in the form of an educated guess or hypothesis. The task then becomes to collect information, which may support, refute, or refine that story.

With a topic as complex as the relationship between wolves, the ecosystems they inhabit, and humans—their primary predator as well as their biggest admirer—the methods of a wildlife tracker come in handy. We will never have all the evidence required to know the exact nature of this relationship. Studies of something as simple as the diet of wolves in a single area can yield different results depending on the specific ecological context and environmental conditions at the time of the study. When we progress to something more complex, such as the cost or value to humans of having wolves in the landscape, the number of variables, which can greatly influence the results you get, grows exponentially. Wildlife trackers are comfortable with ambiguity; it's the nature of their work—indeed, having a sound assessment of what can and cannot be determined from tracks and signs is as much a skill in wildlife tracking as is actually resolving any questions from the evidence. In a society that craves sound-bite explanations and definitive answers, the realization that the story will forever be incomplete, and that our understanding will therefore (if we are paying attention) continue to shift as more information is collected, is the antidote to the polarization of opinions which characterizes the subject of wolves, wildlands, and humans.

While the tracker is comfortable living with ambiguity and with having to amend explanations of the unfolding story, he or she must also be comfortable stepping beyond where the trail disappears, anticipating where it might continue, and striking off in that direction. Finding elusive wildlife such as wolves in most landscapes in the Pacific Northwest absolutely requires going beyond what can be known definitively, making predictions, and testing them. Sometimes we are right, and the tracks appear again in the dust of a game trail. Other times we are wrong and need to make a new prediction about where our quarry has gone and try again. This is an important concept for trying to understand complex systems. While we are still unraveling how ecosystems function, the role wolves play in them, and how we as humans figure in the mixture, there is a great deal of evidence at hand that leads toward some reasonable conclusions. We know that these understandings will change over time, but this shouldn't paralyze us from acting if we hope to advance our understanding or attempt to conserve our world's natural heritage.

An example of the wildlife tracker's methods of discovery comes from a class I taught with my colleague Marcus Reynerson along a river on the west slope of the Washington Cascades. The river canyon, catching the westward-moving weather coming off the Pacific, spends most of the winter blanketed in clouds and perennial drizzle. Wolves have likely been absent from this watershed for a century or more. Leaving our vehicles at a turn-out on the gravel road that runs up the lower section of the canyon, Marcus and I set off with our students into a tangle of young trees and brush but soon entered a beautiful grove of old Sitka spruce and western red-cedar. Elk tracks and scats laced the forest floor, telling us that a small herd had been here not long ago. At the edge of the conifer grove a steep bank led down to the river floodplain. There a sea of forty-foot-tall red alders extended toward the river which bent away from us, wandering east toward the far edge of the valley. As the river had meandered over the years, a dense deciduous forest had slowly occupied the stones and sand left in its wake.

Clumps of willows dotted the shoreline, and signs of elk were apparent at each patch of shrubs. Elk tracks approached each shrub, and a pair of front tracks faced each one, about a foot away from it. At one of these willow clumps, we asked our students to interpret the activity of the elk whose tracks led up to it. Several of them met this question with a quizzical and rigorous inspection of the interior of the footprints, as if some mysterious answer was written in the grains of sand within the tracks themselves. After their initial study, we walked our students through the mystery. We asked "Are these front or hind feet of the elk?" A student returned, "How can you tell the difference between front and hind tracks?" While the front tracks of elk are larger than the hinds, determining which foot was which could also be sorted out with simple deductive reasoning. "If these were the hind feet, where would the rest of the elk be?" I queried. Chuckles emerged as several of them imagined the rest of the elk literally in the willow or the rather unlikely possibility that an elk was walking around on its hind legs. I then asked, "If these are the animal's front feet, where was its head?" Lights come on. "It was feeding!" Indeed, upon inspection, each willow along the river showed the signs of elk browsing: the terminal bud of many branches had been roughly removed by the elk, which, lacking incisors on its upper palate, simply clamps down and yanks off the tip of each branch, leaving a distinctly rough appearance to the wound.

But the most interesting story we discovered that day was tucked away in a small clearing among the thick growth of alders: a line of stunted cottonwood saplings, likely each a sucker from a single root system. The level of attention the deer and elk had given these stunted shoots was readily apparent. Each had a bonsai appearance, with the terminal bud removed and a growth pattern indicating that such heavy browsing had been repetitive over years.

Numerous research projects around western North America have documented the disproportionate impacts of elk and other ungulates (hoofed mammals) on riparian (streamside) vegetation. In the western portions of the Pacific Northwest, black cottonwood, several species of willow, and red alders dominate early successional stages of streamside forests. Of these three fast-growing species, most of the year alder is far less palatable to ungulates because of a higher abundance of tannins and other chemicals in their tissues. The browsing pattern left by the elk we had been following illustrated this preference as they bounced from one willow patch to the next, bypassing the more numerous alders.

So while the alders around these cottonwoods had grown straight and abundantly toward the sky, defining the current forest canopy, the cottonwoods' growth had been retarded, denying them a foothold in the canopy that now loomed high above them. The elk, through their dietary preferences, had helped shape the structure of this riparian forest, encouraging the growth of a monoculture of alders. Decades from now the absence of cottonwoods would be felt by many other species that take advantage of their asymmetrical growth pattern which often produces large crotches and rotting trunk cavities that become homes and nests for everything from flying squirrels to the colonial nesting common mergansers.

What was the composition of riparian forests in this valley before the wolves disappeared? Will the return of wolves to places such as this in the Pacific Northwest alter the course of future riparian forests and the deer and elk that reside there, as it has been suggested they have in the Rocky Mountains? Have human attitudes changed enough to allow wolves to return to places like this river watershed, located less than an hour from the city of Seattle? If so, will wolves find their way back here of their own accord? If they do return, will they find this canyon a desirable place to live, now primarily second and third growth forest, compared to the old growth that existed here when wolves last roamed?

In the chapters that follow, I attempt to shed some light on questions such as these while also illuminating the associated ecological principles and conservation challenges. The results of these relationships have often been written into the very structure of landscapes in the Northwest. This book explores the lives of

Two young wolves explore their surroundings in a meadow on the central coast of British Columbia.

wolves and the ecology of wildlands, in places where wolves survived through the centuries of persecution, in landscapes where they have been absent for decades, and in areas where they have just recently returned.

The plight of the earth's biological heritage in the twenty-first century and beyond will be defined by human choices and by our ability to reckon with our own nature. That we have propelled the biosphere on an unpredictable trajectory is clear, defined by the initiation of large-scale shifts in the planet's climate, the human-caused degradation of just about every ecosystem on Earth, and the extinctions of large percentages of its existent species. Though our collective efforts in recent decades to avert this biological cataclysm appear to have been relatively fruitless, it continues to behoove us to do some damage control and to attempt to create conditions that will support as much biodiversity as possible on the other side of what could be a sixth great extinction event now reshaping life on earth.

Wolves can help us understand the intricacies of this modern dilemma. They cross numerous ecological boundaries and require large expanses of land to survive. They are highly interactive with other parts of the natural world, and their presence increases ecological complexity in many landscapes which in turn tends to create more resilience in the face of disturbances. Thus in ecosystems where apex carnivores, such as wolves, support greater levels of diversity and more complex systems, their presence could be invaluable at times of massive environmental upheavals.

Conversely, wolves are far from a panacea. They, like humans, are driven by biological imperatives for survival. When the landscape around them changes, their impacts change as well, as they strive to meet their needs in the safest and most efficient way possible. In landscapes highly modified by humans, wolves may not be tolerated at levels that initiate the ecological shifts described in areas with more abundant populations. In rare instances, human-caused changes to ecosystems can lead to conditions where wolf populations flourish while the defensive strategies of prey animals are compromised. Human choices and behavior are ultimately responsible in both situations—where wolves serve a role of increasing resilience and integrity in ecosystems and where they are excluded or their ecological relationships altered.

Wolves also reflect our relationship with the world around us. We love them, we hate them, we fear them, we idolize them, we blame them, we are fascinated by them, feelings that mirror our changing relationship with our environment and the evolving ethical guidelines that drive our behavior within it.

For a wildlife tracker or other researcher, making clear, objective observations is vital. Without vigilance on this matter, it is easy for prior experiences and beliefs to cloud one's vision. To this end, I have attempted to critically review as much existing research as possible and to spend time with people of varied experience and with varied orientations toward wolves and wildlands. I have not shied away from topics that are contentious—with wolves, such topics are not hard to find. I have also attempted to experience each issue directly in the field. What I have presented here is the synthesis of this material in as clear and transparent a way as possible. As is often the case when one approaches a subject with an open and curious perspective, over the course of researching this book my own understanding of the role of wolves and humans in the world has shifted and deepened, and my earlier opinions about appropriate behavior by humans as individuals and as a society have similarly evolved.

From the social sciences we've also learned the value in recognizing influences that will inherently color our research. In this way, both the researcher and the reader can reckon consciously with these influences in their interpretation of the results. My own approach has certainly been influenced by my background as a person of European-American descent, my training in biological science and conservation, and my devotion to outdoor adventure, experiential education, and photography. I hope these perspectives will also bring the world of our wild neighbors into sharper focus.

My desire to write this book stemmed from my love of wild landscapes, fascination with the lives of the other creatures that we humans share the world with, and a determination to understand and care for the places that inspire and provide for us. Even as definitive answers to many questions and specific prescriptions for how best to proceed continue to elude us, I hope that readers will share my sense of wonder at the magical and ever-changing character of the Pacific Northwest and its myriad of inhabitants.

My own passion notwithstanding, in the pages that follow it has not been my intention to convince anyone of anything about wolves, wildlands, or people. I've tried to lay out the trail as I discovered it at this moment in time and to provide a window into a fascinating and complex world. What you make of it is up to you.

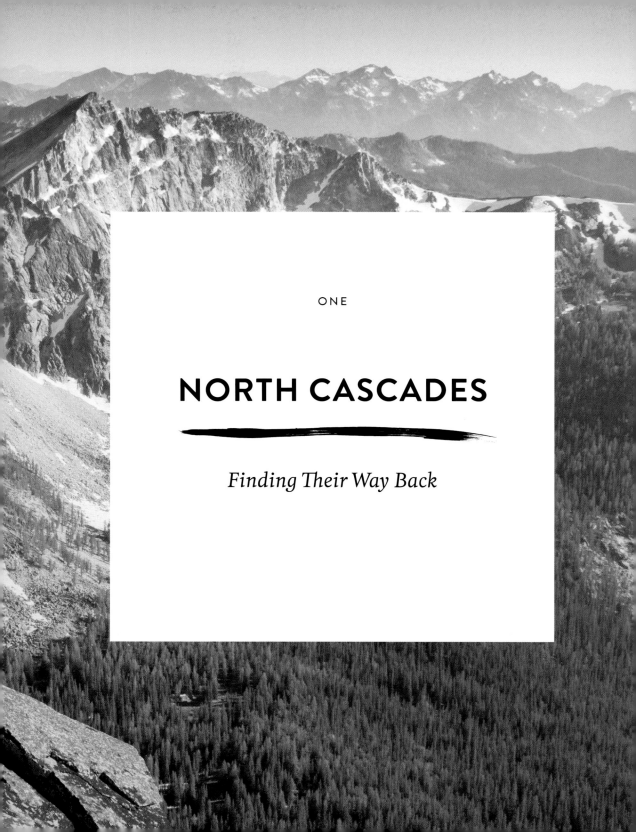

ONE

NORTH CASCADES

Finding Their Way Back

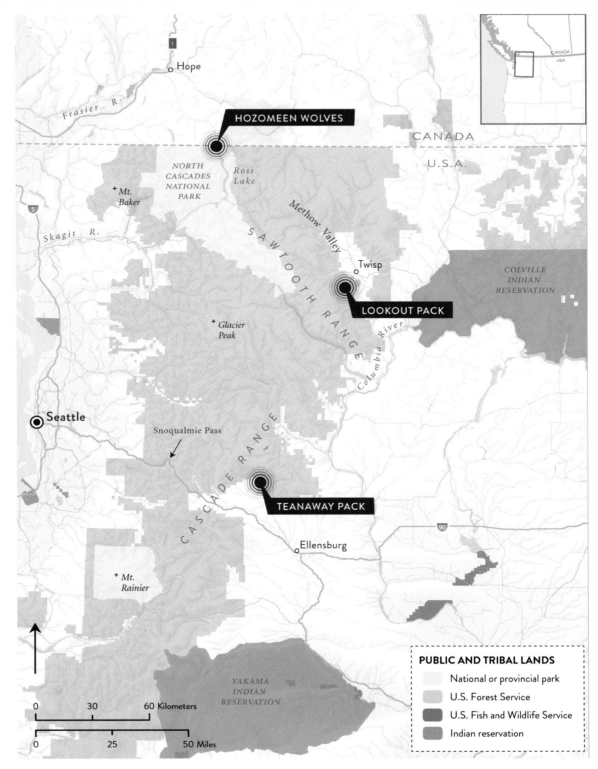

HOZOMEEN WOLVES

CANADA
U.S.A.

NORTH
CASCADES
NATIONAL
PARK

Ross
Lake

*Mt.
Baker*

Frasier R.

Hope

Skagit R.

Methow Valley

Twisp

LOOKOUT PACK

COLVILLE
INDIAN
RESERVATION

*Glacier
Peak*

Columbia River

SAWTOOTH RANGE

Seattle

Snoqualmie Pass

CASCADE RANGE

TEANAWAY PACK

Ellensburg

*Mt.
Rainier*

*YAKAMA
INDIAN
RESERVATION*

PUBLIC AND TRIBAL LANDS

National or provincial park

U.S. Forest Service

U.S. Fish and Wildlife Service

Indian reservation

0   30   60 Kilometers

0   25   50 Miles

NORTH CASCADES

OW DO THEY FIND these places?!" I thought to myself as we picked our way along the trail of a large wolf along a roadless ridgeline on the eastern edge of the North Cascades. After nearly two decades of following the tracks of wild animals through mountains, deserts, and forests from Mexico to Alaska and the Atlantic shores of Maine to the west coast of Vancouver Island, I have become accustomed to wild creatures leading me to amazing hidden places: a caribou trail leading to a secret emerald meadow tucked away in boreal forest, cougar tracks descending through a break in the desert rimrock to a hidden cache of water in a sandstone pothole, the wandering path of a wolverine traversing under massive mountain faces to a small notch on a ragged ridge, or the trail of deer descending from a ridge of burned timber to a deep ravine with towering Douglas firs, ponderosa pines, and a tiny spring bubbling up out of the ground. Even so, each new discovery is a mystery unfolding and perhaps for me the most engaging part of studying the lives of wild creatures in wild places.

It was early May of 2010, and at that point I had lived, worked, and played in this part of Washington on and off for more than ten years. I probably knew the backcountry there as well as or better than any other range of mountains in the Northwest. Yet following the trail of this wolf brought me into a landscape that I didn't think existed anymore. The steep sides of the ridge had protected it from road building, marauding cattle, and timber harvesting. While the higher elevations of these mountains are protected in designated wilderness areas and a national park, the lower elevations are generally awash in a convoluted road system, divided by timber harvest cuts, and overgrazed by cattle under an archaic and broken permitting system. It is a landscape I love but also a landscape defined as much by its scars as by its primal state. Still, as we picked out the scuff marks and occasional clear tracks the wolf had left the night before, I took in the stately, ancient trees, the native wildflowers and grasses, and the notable lack of invasive plant species. Around us stretched an interconnected series of roadless and trailless ridges stretching for several miles, steep sides dropping and defending the pristine ridge tops from the roads, cattle, logging, and failed homesteads that have left their marks across every such landscape in our region.

PREVIOUS: Snow from a summer storm dusts the top of Mount Redoubt in the North Cascades, one of the most rugged mountain ranges in the continental United States.

My traveling companion that day was Brandon Sheely, a life-long hunter and wilderness aficionado. Brandon has covered more miles hiking and exploring around the Methow Valley than anyone else I know and can always be counted on for a lively jaunt in the hills. It was overcast and breezy. Mist hung in the valley bottoms. The high peaks of the Sawtooth Range and Cascade Crest beyond were blanketed in snow. After a strenuous ascent through the forest, we had crested the ridge and come across the trail of a single large wolf along the ridgeline. Brandon, who had tracked the wolves in this area numerous times, surmised from the size of this wolf's tracks that it might be the dominant male from the Lookout pack, whose territory we were traversing. We moved slowly but steadily along the game trail, stopping at blind spots on the ridge, attempting not to spook anything that might be just out of sight. Noting the shifting wind and crossing fresh tracks of deer and bear, we paused to scan the hillsides below for grazing animals.

Picking out scuffs and tracks among the pine duff on the forest floor and between the clumps of bitterbrush, bunchgrass, and balsamroot in the open, we continued to follow the wolf. The animal had passed this part of the ridge after the rain had stopped in the middle of the night. In places the tracks would disappear, but by reading the landscape we could predict its likely route of travel and carry on in that direction until we picked up disturbances again. The wolf's trail cut under a knoll on the ridgeline toward a pass that drops into the next canyon. As we progressed toward the pass ourselves, a long, low, mournful howl ascended from across the valley, out of the mist below. We paused, transfixed, listening. The hair stood up on the back of my neck.

Where did these wolves come from? By most accounts, these mountains have been devoid of their kind for over a century. When was the last time wolves ran this ridgeline? Were these wolves picking out traces of their ancestors here? Brandon and I continued along the trail. The wolf had deposited some urine on a prominent large pine and scraped the duff just beyond it, a common scent-marking behavior and a sign of a resident animal. We carried on along the ridge. Scats and scent-marking spots became common, about every twenty yards or less, a familiar pattern that I've seen near rendezvous or den sites in other areas. After another half mile, the tracks dropped off the ridge to the north into the timber. We did not follow, but the story of the Lookout pack would continue to unfold on the slopes below in the weeks to come.

Two years earlier in the spring of 2008, I had received an email from a colleague at the citizen science wildlife-monitoring project that I co-manage. The email, also

Wolf tracks along a ridge on the eastern edge of the North Cascades in the home range of the Lookout pack.

sent to the project's advisory council, included a photo taken by a remote camera that project volunteers had set up in Okanogan County not far from where Brandon and I were exploring that spring day. Following up on reports of howling and a live-sighting of a group of four wolflike animals in the area, we had recruited and trained local volunteers to set up and monitor a set of motion-sensing cameras in areas the animals were likely frequenting. At the end of May, their efforts yielded a photograph of a large canid in a small and sparsely populated valley in a watershed of the Methow River. Volunteers had baited a rock with a scent lure to attract the animal into the range of the camera. The creature in the image had rubbed against the rock, and volunteers had collected hair samples and large scats containing deer hair and bone fragments, for DNA analysis. Was it a wolf? The email's language reflected hesitancy to make this leap—"large canid" was the term used.

The verdict about the photograph was mixed. Determining a pure wild wolf from an escaped hybrid wolf-dog can be nearly impossible visually, and biologists from the WDFW had been burned by this back in the 1990s when they trapped and radio collared what turned out to be, based on a subsequent DNA analysis,

LEFT: *Springtime on the east slope of the Cascades in the territory of the Lookout pack. The high peaks of Washington's Sawtooth Range rise in the distance.*

BELOW: *This image of wolf pups was taken in 2008 by a motion-sensing camera deployed by Ray Robertson for the Cascades Citizen Wildlife Monitoring Project managed by Conservation Northwest. Images from this set were the first photographs of wolf pups in the state of Washington since the 1930s.*

an escaped semi-domesticated hybrid which had to be recaptured and taken out of the wild. But based on this mounting evidence, the WDFW initiated a live-trapping effort. Eventually genetic analysis from the DNA samples confirmed that, indeed, the animal we had photographed was a wild wolf. Two wolves, the breeding male and female of the Lookout pack, were captured, radio collared, and released. Physical examination of the female revealed she was lactating and had likely given birth to a litter of pups that spring.

Shortly after this another project volunteer, Ray Robertson, set up a camera near what appeared to be the rendezvous site for the pack, capturing photos of multiple wolf pups. This was the first definitive documentation that wolves had reproduced successfully in the state of Washington since the 1930s. Ray continued to monitor the Lookout pack for the United States Forest Service (USFS) as a contractor for the next several years, working in conjunction with the WDFW and United States Fish and Wildlife Service (USFWS).

Genetic analysis of these wolves revealed that the male's DNA linked it to the distinctive genetics of wolves from coastal British Columbia. This exciting find, providing evidence of biological connectivity between the North Cascades and the extensive wildlands farther north in British Columbia, was warmly welcomed by researchers and conservationists involved in wildlife recovery and wildlands conservation across the Northwest.

Glacial ice still clings tenuously to the steep mountain faces in the North Cascades and lingers in shadowed north-facing basins. Dense rainforest cloaks lower-elevation slopes on the western side while the eastern side of the mountains is dominated by arid pine timberlands and open shrub steppe. Rivers draining both east and west are home to salmon runs; these oceanic migrants fight their way up frigid glacier-fed and forest-shaded streams to spawn in the clean gravel of mountain streambeds.

The North Cascades—sometimes called the Alps of America—are the most heavily glaciated mountain range in the continental United States. From Snoqualmie Pass in the center of Washington State, they reach north to the Frasier River in British Columbia, generally becoming taller and more rugged and wild as one travels north. With peaks ranging from seven thousand to nearly eleven thousand feet, the North Cascades are part of the most important geographic boundary in the region, capturing the moisture and moderate maritime temperatures created by the Pacific Ocean on their western slopes and creating an arid rain shadow on the eastern slopes with the more extreme temperatures that characterize continental climates. Extremes also define the enormous relief seen in these mountains with valley bottoms up to eight thousand feet below the peaks.

The North Cascades are also a vital connector of wildlands that stretch down the coast from Alaska to Mexico and inland to the Rocky Mountains. The rest of the Cascade Range runs south through Oregon and into northern California where it merges into the Sierra Nevada. The rolling mountains of the Okanogan and Columbian Highlands to the east stretch to the Selkirk Mountains and the northern Rockies. To the north, the North Cascades merge with the even larger and wilder Coast Range of British Columbia.

With several years of data collected from the two radio-collared wolves of the Lookout pack, as well as other sighting reports, remote camera images, and snow-tracking efforts, a picture of what life is like for wolves on the east slope of the North Cascades has begun to emerge. While much of this picture is predictable and similar to what we know about wolves everywhere, it is highlighted by some uniquely North Cascades vignettes.

Between 2008 and 2010, the Lookout pack occupied a home range of about 355 square miles, spanning elevations from 1500 feet to more than 8000 feet above sea level. The area included arid shrub steppe, dry pine forests, dense subalpine mixed-conifer stands, and even some alpine tundra along the crest of the Sawtooths. The pack numbered as many as nine or ten animals before poaching incidents and dispersals reduced it considerably. In the spring of 2010 the female's radio collar stopped functioning and no pups were documented. By the winter of 2010-2011, Ray Robertson noted reliable reports of only two animals, the alpha male and a second smaller animal, likely one of his offspring, in the pack's territory.

The Lookout pack's use of higher elevations in the late summer is notable. Data collected by the WDFW on the pack's location indicates that, while they made forays to higher elevations occasionally throughout the year, they focused on higher-elevation habitat during the late summer and fall. In contrast, the Rocky Mountain packs tend to linger at lower elevations and in valley-bottom habitat throughout the year. The difference may have to do with the much wider variety of prey species for wolves in the Rockies, including two species of deer, elk, and moose, while the Lookout pack's seasonal migration is in part tied to their single primary food source, mule deer (*Odocoileus hemionus*), many of which also migrate to higher elevations for the summer.

While they are up along the Sawtooth Crest enjoying long days and subalpine meadows, they also find some tasty mountain treats, as when a hiker in the area observed a Lookout pack wolf eating a hoary marmot (*Marmota caligata*). Back at lower elevations I have found signs of wolves eating not just deer but also beavers and wild turkeys. Ray Robertson reported following the tracks of several Lookout wolves who had followed the trail of a moose—rarely seen wandering through their home range—apparently hunting it.

The impact of wolves and other carnivores on not just the numbers of their prey but also on the behavior of prey has been the focus of a great deal of study with the return of wolves to the western United States in recent decades. Driving to work one morning in April 2008, Ray observed a herd of ten to fifteen deer in a field behaving in a manner which caught his attention. "Slowing down, I saw that the group was so close together they were almost touching each other as they trotted along changing directions often in unison," he recounted. "My first thought was how odd this was to see that many deer acting like they were being herded by something. Stopping the car, I looked behind them to where the flat field turned into a steep shrub-steppe slope and saw two wolves looking at the deer and looking at me."

The wolves soon disappeared into the brush. Ray reflected that he couldn't recall seeing deer behave in quite that way in the valley before. Bunching and travel in unison, such as Ray observed, is classic social defensive behavior for hoofed animals dealing with a running predator such as a wolf. For Ray, a long-time resident of a valley where deer are more ubiquitous than squirrels, this behavior in deer in the presence of wolves was striking.

Perhaps the most intriguing anecdotal observation of the Lookout pack's diet thus far comes from John Rohrer, a wildlife biologist for the Okanogan-Wenatchee National Forest and one of the scientists responsible for monitoring these wolves. What he discovered hints at what historically may have been, and might become again, a typical scene along salmon-bearing waterways in the North Cascades.

Rohrer recounted following up, with two colleagues, on a report of a wolf in the Twisp River valley in late November 2009. They confirmed that both radio-collared animals were there, but the story didn't end at that. "There was fresh snow on the ground and there was a maze of wolf tracks. It appeared a number of wolves had been in the immediate area for at least a couple days. We assumed they had made a kill and started following tracks assuming we would find the remains of a deer or moose carcass. Instead what we found, in a dense patch of riparian shrubs, was the bony remnants of large fish, which evidently had been fed on by the wolves. It also looked like at least one or two wolves had been bedded right there adjacent to the remains."

When Rohrer got out of the field he did a little research. "Our district fish biologists were very skeptical about there being live or even spawned-out salmon carcasses in that area in late November. I made a few phone calls to local hatcheries and found that salmon carcasses had been dumped off one of the bridges in the Twisp River watershed that fall. The wolves were making good use of a scarce and seasonal food source."

Such deposits of salmon carcasses from hatcheries are part of restoration efforts for struggling salmon runs in the Pacific Northwest. Putting salmon carcasses into streams has been shown to increase the survival of juvenile salmon, possibly because of the increased nutrients available from the decaying fish. Apparently the salmon are not the only beneficiaries of this program—fitting, given the tremendous ecological value salmon have provided for a wide range of species in our region.

Salmon are not typically the first thing people think of when they imagine what wolves eat. There are no published accounts of wolves eating fish in the Rocky Mountains to the east. The genetics of the breeding male from the Lookout pack link it to the wolves of coastal British Columbia where salmon consumption is a mainstay of their diet at some times of the year. Did these wolves bring an awareness of this food source with them, or, as opportunistic foragers, did they just seize on an easy resource? As wolves reoccupy areas from which they have long been absent, we may discover a great deal about the still murky relationship between these two iconic species in the Pacific Northwest.

WHEN WILDERNESS
CAN'T PROTECT

Since the discovery of the Lookout pack in 2008, the recovery of wolves in the North Cascades and elsewhere in the Pacific Northwest has taken unexpected twists and turns. During the fall of 2010, our citizen science project helped document another wolf pack at the southern end of the North Cascades ecosystem. Genetically related to the Lookout pack, the Teanaway pack occupies a territory just north of Interstate 90, basically in the geographic center of the state, much farther south than anyone anticipated the second confirmed pack in the North Cascades would appear.

Wolves' wide-ranging nature and their role as a pivotal carnivore in a complex web of ecological interactions make them an ideal species for helping us to understand many intricacies in our region's ecosystems, including the changing landscape of our society in relationship to them. Wolves are a lot like humans. We overlap in our social nature, in many landscapes that we like to use, and in several foods that we enjoy. As we look into this mirror, wolves are also good at pointing out how our choices affect the region's ecosystems as well.

The fallout from overlapping human and wolf habitats in the North Cascades would soon be felt, especially by the Lookout pack. It does not overstate the case to say that a major social schism divides the Pacific Northwest (and indeed western civilization) concerning how we value wildlands and wildlife. Wolves have an uncanny knack for exposing this fault line in our society. With hundreds of square miles of remote wilderness to choose from, the pack's breeding female dug her den adjacent to private ranchland just a few miles from the town of Twisp in Washington. In the three years following their discovery, there was no documented depredation on livestock by members of the Lookout pack, but just their presence in such a landscape created a charged atmosphere, and three different poaching events were documented during this time.

An opinion poll conducted statewide in Washington found that the clear majority of people thought favorably of wolves and believed they should be protected in the state. However this study suggests, and other similar studies show, that

This pair of glacier-carved lakes in the North Cascades is part of the summer range of the Lookout pack.

there is far from a consensus on this issue. People most likely to have an unfavorable view of wolves are rural residents and those involved with the livestock business.

One case of poaching, which ended with multiple state and federal convictions in 2011 and 2012, concerns a landowner within the home range of the Lookout pack. This case dramatizes the fact that the Lookout wolves moved in with some unfriendly neighbors. Why would they do this?

Their choice of home reflects a downside of a generation of conservation initiatives relating to wildlands protection. While large tracts of the North Cascades are federally protected in roadless wilderness area, most of this is high elevation and, while beautiful, totally unrepresentative of the wide variety of critical habitats in the ecosystem. Low elevations are significantly underrepresented—lands

that are critical winter habitat for deer, elk, and many other species. Wolves go where there is food. We are learning that, for wolves on the east side of the North Cascades, this often means leaving the official wilderness and roadless areas to go down into low-elevation landscapes. Protecting remote, rugged, and high-elevation landscapes was the task of the day for decades; "Saving the last best places" was the motto. But as the field of conservation biology has matured we have realized that islands of protected lands, even large islands like the core of the North Cascades, are not self-contained ecosystems. Without protection of adjacent land-scapes and connectivity with other functioning wild landscapes, the ecological effectiveness of these islands, while scenic for human entertainment, can become greatly diminished as wider-ranging animal species have to leave for part of the year to get to the important habitat outside of preserved areas. Those who don't travel more widely might be cut off from other populations and become suscep-tible to in-breeding or local extinction from disease or particularly difficult sea-sonal conditions.

Conversely, the discovery of the Lookout wolves in the Methow Valley demon-strated the continued functioning connectivity of the North Cascades with other far-flung wildlands, an important sign of health and vigor in this ecosystem. That they chose to den in an area that is prime winter and spring mule deer habitat is predictable. Most of this habitat lies within a matrix of state, federal, and pri-vate lands and is much less protected than higher-elevation, less biologically pro-ductive, areas. This ecological irony is typical of landscapes across the region and indeed around the world. Humans, like wolves, know where productive lands are and claim them for themselves.

Once summer arrives, many of the mule deer from the part of the Methow River watershed where the Lookout pack established itself move up to higher ele-vations. Here the Sawtooth Range cuts a long swath south and east from the core of the North Cascades out toward the Columbia River. Sitting high above the Methow Valley to the north and Lake Chelan to the south, the high Sawtooths are defined by jagged, glacially carved ridges and peaks, deep cirques, and U-shaped valleys (thus the name Sawtooth), much of this contained within the Lake Chelan Sawtooth Wil-derness. The wolves, following the deer up in elevation in the late summer, make use of the remote roadless landscapes that we have so diligently set aside for such uses. But during the most sensitive time of the year for wolves (pup rearing) and, not coincidentally, livestock operations (calving), wolves on the eastern slope of the North Cascades require low-elevation habitat. This makes them far more likely to come in contact with livestock and human activities.

A very different story, however, is unfolding farther west and deeper into the core of the North Cascades ecosystem. As the Lookout pack was establishing itself on the periphery of the North Cascades, signs of wolves were beginning to pop up again deep in one of the most remote and rugged sections of these mountains.

THE HOZOMEEN WOLVES

Wolf tracks in the mud, large and nearly unmistakable—I smiled at the discovery and began to follow the trail. I moved along what I anticipated was the route the wolf had traveled across the lakebed, occasionally picking up a track in the mud between stretches of tangled driftwood. Ross Lake, a twenty-three-mile-long reservoir, was created by a dam completed on the Skagit River in 1949. The reservoir is long and thin, filling the bottom of a deep valley flanked by rugged, remote mountains, including the striking rocky summit of Hozomeen Mountain at the north end of the reservoir. The prevailing west winds come barreling up the valley from where the Skagit River leaves the Cascades and drains into Puget Sound, turning north as the river valley bends where Ross Lake begins. Branches and logs get washed into the reservoir from feeder streams, and the wind drives them north toward the top of the lake. The reservoir, managed by an electrical utility company, generates electricity for the city of Seattle; lake levels fluctuate depending on anticipated needs for electricity.

When I discovered the tracks during the late spring of 2011, the water level had been exceptionally low for months, creating a vast open savannah about two miles wide and long at the north end of the reservoir. This open landscape was punctuated by old stumps of trees that had been cut before the reservoir began flooding the valley decades earlier. I had spotted several black-tailed deer (*Odocoileus hemionus columbianus*) earlier in the day, and abundant tracks and feeding sign indicated that they were using the open area extensively. Wolves, drawn to openings in forested environments and to deer, had apparently also been out exploring—but deer were not what they had found.

A few hundred yards from where I first picked up the tracks I found several clumps of fur in a patch of open ground with a number of wolf tracks around it. Picking up the dark wavy hair made it clear to me that the animal this fur had belonged to was likely now deceased and that it was not a deer. Looking around, I pondered where the rest of the carcass might be. In open areas such as this wolves often make their kills on the edge of the trees or just inside the forest. I looked to the forest edge and noted a small alcove where a stream fed into the currently dry lake. The black bear carcass was there, right beside the treeline where the flat valley bottom turned upward and the trees begin above the high-water line of the reservoir. The carcass had been there for some time and had been reduced mainly to bones and tufts of hair.

The condition of the teeth suggested that this bear was quite old. One of its canine teeth was fractured, and the other well rounded. All of its lower incisors were missing, and the bone had healed over, indicating that they had been missing for some time. Black bears in the North Cascades eat a lot of inner bark from conifer trees, craving the sugar content of the sap flowing through it. They pull the outer bark off with their curved claws and then scrape the cambium from the trunk with their lower incisors. Many red-cedar trees in the valley bore the telltale signs of this feeding behavior. I had heard that this will cause their incisors to rot but had never before seen evidence of it myself.

The previous day, a few miles north of where the carcass lay and on the other side of the international border, I had found a wolf scat filled with the same wavy dark hair. Weeks later I talked with Paul Frame, a biologist working on wolves for the WDFW. He too had come across the carcass, about a month before I had, after finding a wolf scat several miles to the east that had a bear claw and lots of fur in it. While the scats indicated that a wolf had fed on the bear, the remains were too far gone to determine whether wolves had killed it or if it just curled up and died before being discovered by a wolf or wolves. I found parts of all four limbs. The entire rib cage was there as were the skull and mandible. Finding the entire carcass in one spot is unusual but not unheard of at carcasses that have been fed on by a pack of wolves. Wolves, social carnivores, often carry off pieces of the carcass to feed at some distance from the kill away from pack mates. Had a lone wolf visited the carcass, or had multiple animals found it after it was long dead and only nibbled around the edges? How often do bears end up on the menu for wolves here in the North Cascades? As with most discoveries like this, more questions than answers result.

Ross Lake sits at a geological, ecological, and political crossroads. It is located in a deep valley near the middle of the North Cascades, a range whose geological creation at the edge of North America has been a turbulent process. Over millions of years several hundred chunks of seafloor plastered onto the continent as volcanic events twisted, shuffled, and uplifted the region while the Pacific Ocean tectonic plate subducted under the North American plate. Most recently the ice ages left their marks here by gouging out deep U-shaped valleys, polishing massive cliffs along the canyon walls, and carving soaring mountain ridges and peaks. The reservoir's striking emerald green waters are a product of the glacial silt that gets funneled into it from the numerous active glaciers which still cling to the mountain tops that rise over seven thousand feet above the lake within just a few miles of the shore.

Technically the reservoir is west of the Cascade Crest, which divides the west and east sides of the range. However, Mount Baker and the high peaks of the Picket Range, to the west of the reservoir, leave the valley in a rain shadow. The forests west of the valley are typical of the rainforests that define the Pacific Northwest—groves of immense ancient western red-cedars line the stream drainages, and dark stands of Pacific silver fir and western hemlock creep up the mountainsides. But decreased precipitation in the valley bottom creates conditions closer to those in the more arid landscapes on the east side of the Cascades. Stately ponderosa pines dot the meadows on the north end of the lake, and the eastern side of the valley marks the beginning of the arid interior forests that define the eastern slope of the Cascades.

The reservoir's surface sits at about 1600 feet above sea level, a welcome, low-elevation refuge from the much harsher winter conditions higher in the mountains all around. The snowpack is often transient in the winter, in contrast to the record-breaking amounts that amass on the surrounding mountains (Mount Baker set a world record of ninety-six feet for single-year snow accumulation). As on the east side of the Cascades, deer that summer in the high mountains seek out this low-elevation habitat to survive the winters. Apparently the wolves follow.

The reservoir spans the political border between Canada and the United States. Almost all of its U.S. watershed is protected in national forest and national park wilderness. Its Canadian watershed is a matrix of provincial parks and lands managed for timber harvest, marked by a substantial logging road system. Access to the reservoir is limited. Although it is primarily in the United States, the only road access is through Canada at the far northern end of the lake. While the occasional motorboat zips up and down during the summer months (when there is water at the boat launches at the north end), most of the time the only human traffic on the lake are canoeists and backpackers. When the wolves I was tracking are in the United States, they are considered an endangered species at both the federal and state levels and receive the highest level of legal protection available. When they cross the border, which is defined by a swath of trees cut along the forty-ninth parallel, they face a hunting and trapping season that is both lenient and long.

The area around Ross Lake is a wild place where numerous rare carnivores have been seen returning to the North Cascades. North Cascades grizzly bears are occasionally spotted from the lake. History was made in the North Cascades when biologists Scott Fitkin and John Almack documented a pack of wolves with pups during a howling survey at the north end of the lake in 1990. Numerous sightings followed in various parts of the range, including documentation of another pack with pups in the Glacier Peak Wilderness in 1991, about sixty miles to the south of the animals on Ross Lake. These discoveries initiated a flurry of activity and speculation about the recovery of wolves in the North Cascades and beyond. But then the trail went cold. Occasional sightings of wolves trickled in over the next decade, but no reproducing packs were discovered. More than twenty years after that initial confirmation of wolves with pups in the shadow of Hozomeen Mountain on the north end of Ross

Lake, the activities of the wolves whose tracks I had followed to a bear carcass in that same area in the spring of 2011 remain a mystery.

THE TRIALS OF THE LOOKOUT PACK

Wolf den sites often have an ethereal, mythical quality to them. Usually located in some secluded or overlooked nook in the landscape, a den is typically a large burrow dug into the earth, often at the base of a large tree, stump, or fallen log. Peering into the dark, portal-like hole that disappears into the shadowed depths of the earth, you are witness to the most hidden treasure of a creature of whom so many stories have been told. This private, guarded chamber is the birthplace of a force of nature—friend or foe of humans, depending on the myth—that our species has alternately honored and reviled across the globe and across generations.

One of several burrow entrances at the den site of the Lookout pack on the eastern slope of the North Cascades.

Two years after the confirmation of its existence, something had gone terribly wrong for the Lookout pack right in the middle of denning season. In May of 2010, shortly after I had been exploring their home range with Brandon Sheely, the breeding female disappeared. Biologists were monitoring her almost daily by means of her radio collar as well as with a spotting scope from a vantage several miles from the den site. All had been normal for that time of year: she and other pack members lingered around the den. Based on the activity pattern and season, it appeared likely that she had given birth to a litter of pups. Then one day the collar stopped transmitting and activity at the den ceased.

On a bright day in late May, I joined Ray Robertson, Dan Russell, and several USFS biologists on an investigatory trip to the den site. Dan was working toward his WDFW Master Hunter Permit at the time, and his assistance with this outing was part of the wildlife-related volunteer hours required for this certification. Previously Dan had worked with Ray carrying out monitoring activities for the WDFW and USFS and knew the area and these wolves well. Ray was carrying a remote camera to set up at the den site to attempt to collect information and sort out what had happened to the pack.

The day was pleasant; scattered clouds and a light breeze kept the oncoming summer heat at bay. Hillsides, which a few weeks before had been a riot of yellow with blooming balsamroots, were quieter now, most of the sunflowers having gone to seed. The den site, guarded by a steep ridge several thousand feet tall, took several hours to reach from the closest road, a small gated Forest Service road. We ascended through south-facing bitterbrush steppe and open stands of ponderosa pine and Douglas fir. We passed bountiful small digs where yellow pine chipmunks had been unearthing western spring beauty (*Claytonia lanceolata*) bulbs, a crop also harvested by grizzly bears. Though grizzlies haven't been seen in this part of the North Cascades for many decades, the area is in the federally designated recovery zone for them. The spring beauty diggings made me wonder when they might be back, and how the presence of wolves in the mountains might influence their recovery.

We carried on up the slope, past scattered clumps of serviceberry and a small grove of aspens in a ravine slicing down the ridge. We noted the tracks and scats of black bear, mule deer, and coyote but no sign of the wolves. As we climbed, between breaths, we discussed wolf use of the area and past encounters with wolves and other wildlife, traded notes on the plants we encountered, and stopped to listen to the songs of the neotropical migrant birds that had begun nesting in the area.

Making the ridgeline took nearly two hours for the six of us. Once there we took a break as Ray checked on the remote camera he had set a couple weeks before, swapping out the batteries and memory card and leaving new ones in their place. The camera looks across a ridgeline game trail where we noticed a few fresh wolf tracks in a patch of dust. Before moving toward the den site, we wanted to be as sure as possible that there weren't any wolves present that we would disturb, so Dan and Ray assembled the radio telemetry equipment. Ray picked up the signal of the male close by but coming from a different direction than the den site that was now almost directly below us on the other side of the ridge. As had been the case for several weeks, there was no sound on the frequency of the female.

We dropped off the north side of the ridge and descended to a bench with a good view to break for lunch. Checking the hillside below and the ridge across the valley with binoculars, Jesse McCarty, a Forest Service biologist, noticed a black bear grazing on a south-facing slope across the canyon. We watched it munch on balsamroot flowerheads and meander up and over the far ridge. Before we set off again, Ray checked the radio signal of the male again, picking it up to the west of our location and the den. We continued our descent.

Through scattered ponderosa pines and Douglas firs, the throw mound of the burrow—light subsoil excavated by the wolves and left in a pile next to the den entrance—came into view among the pine grass and lupine. As we approached, I noticed a prominent pine twenty yards from the den with an oval depression at its base—a wolf bed—and beside it the foreleg of a deer, the humerus cracked with the top half missing. The burrow had multiple entrances. Around it scats, bones, and bone fragments littered the ground. Numerous other beds were scattered around under the trees, along a fallen log, and out in a small clearing under a bitterbrush. Several of the scats were small, typical for pup scats. Around the main burrow the ground was bare from the traffic of wolves coming and going at the den. The large throw mound was flattened from weather and traffic as well.

As Ray and Dan selected a tree for setting up the camera, the rest of us explored the area looking for fresh signs of wolves and anything else that might give a clue as to what happened to the missing female. As I peered into a secondary entrance to the den, a rattlesnake emerged from a small excavation. Startled by our presence, it retreated into the den entrance. The entrance diameter of the main burrow was large enough to accommodate a small person, and I shivered at the thought of running into the snake in a confined space like that. None of the scats looked particularly fresh, nor did any of the other sign. It was clear that there were no pups in the area.

What exactly happened to the female and her pups has remained a mystery, though the evidence at hand strongly suggests she was intentionally killed and that the pups died as a result, or were possibly killed as well. If her radio collar had simply stopped transmitting, this would not have affected activity around the den. If she had died of natural causes, the radio collar would have sent out a mortality signal (indicating the collar is no longer moving at all). Given that she disappeared and her collar stopped transmitting anything makes it very likely that she was killed and that the collar was intentionally destroyed. Who might have done such a thing? A number of people who lived in the area knew the general whereabouts of the den and some of them were outspokenly anti-wolf. Whoever carried out the act has done a good job of keeping it to themselves.

The same cannot be said of other poaching incidents related to the Lookout pack. In what will likely remain the most notorious wolf-poaching case in the Pacific Northwest for years to come, several members of a local family with property within the Lookout pack's home range, in an area commonly used during the denning and pup-rearing season, were convicted in the killing of five wolves and in attempting to send the pelt of one to Canada to be certified as having been legally killed there. What makes this case so incredible is how the poachers were caught. In early 2009 they attempted to ship the raw pelt, via FedEx, to an outfitter in Canada who was to have the skin certified as having been killed in Alberta. The package was accepted by the FedEx outlet (at the Walmart in Omak, Washington). However, when the FedEx driver refused to pick up the package because it was leaking something that resembled blood, local law enforcement was contacted. The woman who had brought in the package left a false name and contact information, so some careful investigation ensued. Eventually all the pieces came together, leading officials to a family living within the home range of the Lookout pack who had publically opposed the reestablishment of wolves in the region. Further evidence gained from a search warrant indicated that the family had been involved in shooting at least one wolf and trapping another. This evidence included trophy photographs of a family member posing with one of the dead wolves. It also came to light that the family had been hunting bears and cougars out of season and hunting with hounds, which is also illegal in the state.

When the remote camera was set we retreated back up the way we had come, climbing steadily to reduce our impact on the location. Once on the ridge we dropped off the other side, stopped, and sat in the sun, taking a breather after the climb. Jesse, a student at the University of Washington at the time, recounted some of his grandfather's experiences working as a predator control agent for the federal government in the mid 1900s. The federal government's predator control program was in large part responsible for the final extirpation of wolves from the western United States. Jesse's grandfather had told him about shooting wolves and bears from helicopters, explaining how they would fly the helicopter toward the animal while facing sideways so that he could get a clear shot at the animal. His grandfather was in high demand as he could shoot ambidextrously out either side of the helicopter, often firing over the pilot's shoulder as the helicopter approached its target. Jesse, who was involved in one of the most rigorous, cutting-edge wildlife biology programs in the nation, had told his grandfather how the field of wildlife science has a very different understanding of the role of predators in ecosystems than it did a few decades ago. His grandfather held firm to his beliefs, though, declaring, "That's just college talk; the only good wolf is a dead one."

THE TRAIL AHEAD

With plant communities that span from rainforests to tundra to semi-arid shrub steppe, the North Cascades is a varied and fascinating ecosystem. The wolves reinhabiting this diverse landscape are etching out a new story in the fabric of these mountains. While this story has echoes of the Rockies to the east and the coast of British Columbia to the north, the lives wolves are making for themselves here is distinctively North Cascades.

We can also see that wolves' relationships with people will be varied across the ecosystem as well. It wasn't until four years after the discovery of the Lookout pack that a probable case of livestock depredation was documented. Conversely humans killed numerous wolves from the pack in at least three poaching incidents. Local newspapers in the region have been rife with letters to the editor since the pack's discovery, expressing many points of view and opinions, all of them strongly held. Meanwhile wolves in North Cascades National Park have been carving out a territory so isolated from humans that in almost two decades we have yet to get a firm grasp on what they are up to there. The discovery of the Teanaway pack on the far southern tip of the ecosystem means that wild wolves are about a two-hour drive from downtown Seattle, an exciting development for many in one of the most outdoor-oriented and environmentally engaged cities in the world.

The initial anecdotal observations of professional biologists and citizens alike hint at things even deeper than the way wolves are re-adapting themselves to the North Cascades. These observations begin to get at how the mountains themselves may be changed by these engaging carnivores. Subtle changes in the behavior of deer, signs of feeding on salmon carcasses, and questions about how wolves are relating to other large carnivores in the North Cascades are all intriguing. In order to parse the meaning of these observations, however, we need a greater understanding of wolves as a species and of their ecological role in the Pacific Northwest.

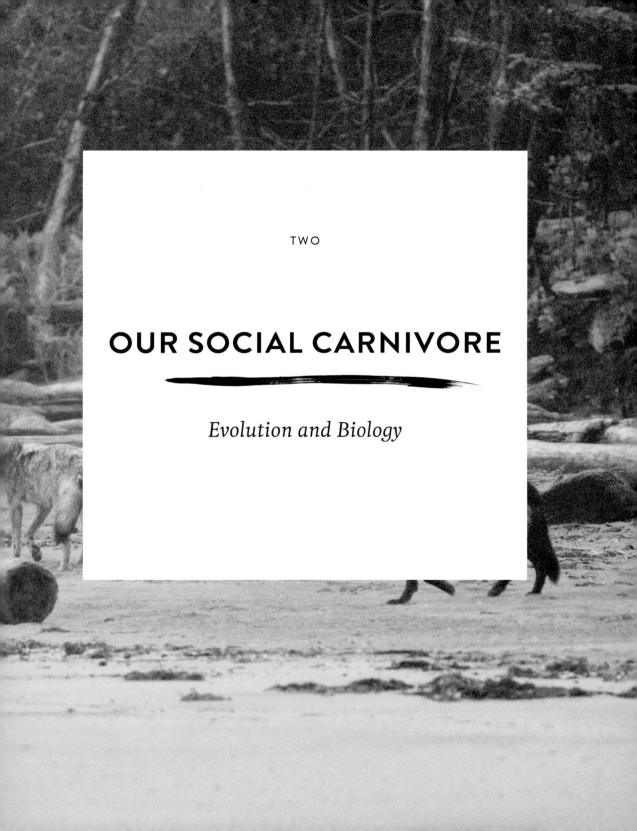

TWO

OUR SOCIAL CARNIVORE

Evolution and Biology

OUR WOLF, *Canis lupus*, was born out of an age of oversized creatures, large mammals that would seem like fairytale characters to us now—extra big, extra furry, with extra big teeth or tusks. Where now it occupies the position of apex carnivore among large mammals in our region, less than fifteen thousand years ago the wolf was competing with even larger beasts for the distinction: American lions (*Panthera leo atrox*, a larger subspecies of the modern African lion), saber-toothed cats (*Smilodon* species), and dire wolves (*Canis dirus*). These species went extinct along with a host of large herbivores (like a two-hundred-pound beaver), leaving *Canis lupus* as the dominant carnivore over most of North America including all of the Pacific Northwest.

Like humans, wolves are a social animal. Every aspect of a wolf's life revolves around its relationship to other wolves. The primary unit of wolf society is the pack, an extended family who travel, hunt, feed, sleep, and play together. This is an unusual situation for large carnivores, most of which are primarily solitary, with adults tolerating each other's presence during the breeding season and avoiding contact most of the rest of the time. Sociability in carnivores is not unheard of, however. Africa has several highly social large carnivores: hyenas, lions, and the African wild dog. In fact, the wolf's distinction as the only social large carnivore in the Pacific Northwest was earned through perseverance and attrition. As other social carnivores, including other large canids, disappeared from our region over the past several million years, most recently the larger dire wolf at the end of the last ice age, the wolf has carried on.

Wolves represent the canid family's most recent evolutionary experiment with highly social big-game hunters. The wolf's closest living relative is the coyote (*Canis latrans*), the two species having parted ways about one and a half million years ago. The now extinct species *Canis lepophagus*, known only from the fossil record, is thought to be the last common ancestor of both. Coyotes are more similar morphologically (in form and structure) and likely behaviorally to *C. lepophagus* than wolves are. While coyotes have carried on with a smaller build, omnivorous diet, and flexible social structure depending on their environment, wolves represent a specific evolutionary change toward a highly social hunter of large game, as we see in their pack structure and their focus on large hoofed mammals as their primary food source. This pattern of evolution from a smaller omnivorous, foxlike creature toward a larger hyper-carnivorous animal has occurred on multiple occasions in the evolutionary history of the canid family.

The dire wolf evolved and lived in North America alongside the gray wolf until the dire wolf's extinction about eight thousand years ago. Between about fourteen thousand years ago and the time when the dire wolf went extinct, at the end of the last period of glaciation on the continent, North America lost about half of its large mammal species. Dire wolves likely occupied all of the Pacific Northwest at one time, and fossil specimens have been collected from Oregon. The dire wolf was larger than our modern gray wolf, with a massive head, proportionately larger teeth, and shorter limbs. Its disappearance left the gray wolf as the sole surviving large social carnivore in the Pacific Northwest in modern times. Two theories exist as to the cause of the dire wolf's extinction: loss of its megafauna prey-base during the Pleistocene extinctions or an inability to compete with *Canis lupus* for the smaller, fleeter prey that remained.

FORM
AND FUNCTION

The gray wolf is the largest living member of the canid family (Canidae) in the world, at an average eighty pounds for females and ninety pounds for males. It is the third largest carnivore (order Carnivora) in the Pacific Northwest, after the grizzly bear (*Ursus arctos*), which is often well over 300 pounds, and the black bear (*Ursus americanus*) at 150 pounds or more, and is similar in size to the mountain lion (*Puma concolor*).

As with many wide-ranging species, wolves tend to be larger the farther north you go and smaller the farther south. This trend is believed to be a result of increased efficiency in conserving heat (larger bodies lose heat more slowly) in colder northern latitudes and shedding heat in warmer southern locations. Bergmann's rule, first articulated in the 1800s by German scientist Christian Bergmann, states that animals within a single species will be on average more massive in colder environments than in warmer conditions. Latitude has often been used as a proxy for studying this effect, as average temperatures decrease as one moves from the equator toward the poles. Studies have confirmed that in general this rule accurately described the majority of mammal species with large geographic distributions, with large mammals, including wolves, being more prone to fit the model than smaller ones.

At an awe-inspiring five and a half feet from nose to tip of tail for males, wolves in the Pacific Northwest are predictably about average in size compared to the distinctly larger wolves of Alaska and northern Canada and smaller wolves of the desert Southwest and Mexico. As with many carnivores, wolves are sexually dimorphic—males are slightly larger than females in size and weight, though are otherwise similar in appearance.

PREVIOUS: *Three wolf pups in the Great Bear Rainforest, British Columbia.*

COYOTE

WOLF

However, even within the Pacific Northwest, wolves vary in size, a seemingly simple fact that has been at the core of recent controversy. In general British Columbia's coastal wolves average slightly smaller than wolves from the Rocky Mountains and interior. Similarly, wolves in the North Cascades in Washington appear to be slightly smaller than wolves farther east in the region, extrapolating from a limited WDFW data sample. Though reliable data is minimal on the size and weight of historic wolves in the region, wolves reintroduced to central Idaho from farther north in the Rockies were probably slightly larger than the wolves that humans extirpated from Idaho in the past, as would be predicted for a population originating from farther north geographically.

In addition to their overall larger size, wolves can be distinguished from coyotes by a broader and deeper snout with a larger nose pad, relatively shorter and more rounded ears, and relatively larger feet.

This is the kernel of truth that spurred a great deal of hyperbole about how the wolves reintroduced into Idaho are much bigger, travel in larger groups, and are more aggressive and dangerous than the original population of wolves from that area. There is no evidence—from historical data or modern behavior of wolves in either location—to support these other statements. Yet the stories have spilled over

COYOTE

WOLF

into Oregon as descendants of this reintroduction have made their way into the northeastern part of the state.

Wolves have a trim body, long legs, elongated skull, and bushy tail, and are adapted for a lifestyle that involves covering long distances on a regular basis. Identifying a wolf in the field can be challenging because of similarities between wolves, coyotes, and some domestic dog breeds. Wolves are also well adapted for cold climates and snow, with their relatively large feet, dense pelage (coat), blockier muzzle, and shorter, more rounded ears than coyotes, a species generally believed to be adapted to more arid environments. Wolves' large feet also aid in swimming, something they do a lot of in parts of our region. While adult wolves are distinctly larger than coyotes and most breeds of dogs, juvenile wolves can be similar in size to coyotes and numerous types of dogs.

Despite the common name *gray wolf*, coloration can vary, including black, which is common in parts of the Northwest, or very pale gray to white, which is less common in the region. Animals that are predominantly gray can also have ruddy orange or reddish highlights, common in coastal populations in British Columbia. All breeds of the domestic dog (*Canis lupus familiaris*) are descendants from wolves and share many of the anatomical aspects of wolves discussed here.

BELOW: Many wolves from British Columbia's coastal population have reddish highlights in the pelage, as seen in this large wolf from the central coast. Its broad head and large stature indicate this is likely a male.

OPPOSITE ABOVE: Wolves that are predominantly black, such as this one from northwestern Montana, can be found across the Pacific Northwest.

OPPOSITE BELOW: This coyote shows the typical coloration for the species, as well as the long pointed ears and narrow pointed snout with a small nose pad.

SKULL AND TEETH: PERCEPTION AND DIET

The skull of a wolf reveals much about the evolutionary history, lifestyle, and biology of this exceptionally successful mammalian carnivore. A skull provides clues about how an animal perceives the world through its senses of sight, smell, and hearing, while the structure of the teeth, jaw, and related elements also offers clues about diet.

The long snout, or rostrum, creates a large nasal cavity, conducive to the excellent sense of smell in wolves and other canines. The relative length of the skull in wolves is greater than in members of the cat (*Felidae*) and weasel (*Mustelidae*) families. This increased length gives wolves less leverage on their canine teeth than these other predators have relative to their size. However, it also makes more room for their highly effective noses which are far more powerful than those of felines or mustelids. A wolf's nasal cavity is many times larger than a human's; their sense of smell has been estimated to be one hundred to ten thousand times better than ours. Besides detecting prey, wolves use their powerful sense of smell for communication with one another. They use a variety of scent glands on their body to communicate both directly and through marking behavior on the landscape.

Note the wolf's large canine and carnasial teeth used for gripping prey (canines) and shearing flesh and bones (carnasials). The well-developed temporomandibular joint allows for a great deal of power in the mandible while the enlarged sagital crest provides a large attachment point for the muscles that contract the jaw. The long rostrum houses the wolf's very sensitive olfactory capabilities.

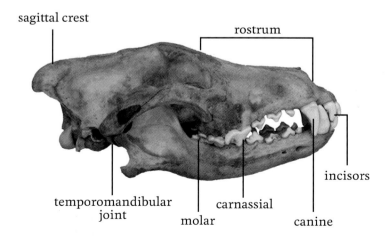

sagittal crest

rostrum

temporomandibular
joint

molar

carnassial

canine

incisors

SIDE VIEW

FRONT VIEW

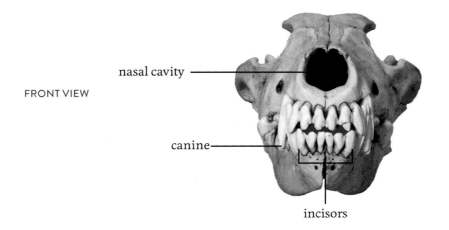

nasal cavity

canine

incisors

BOTTOM VIEW

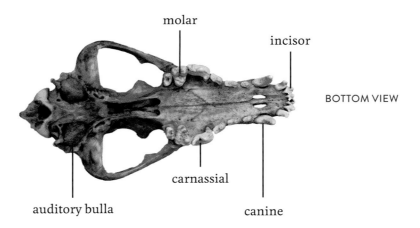

molar

incisor

carnassial

auditory bulla

canine

MANDIBLE

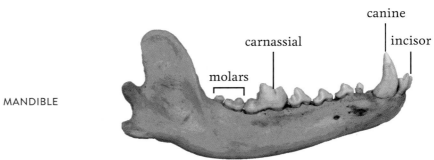

canine

incisor

carnassial

molars

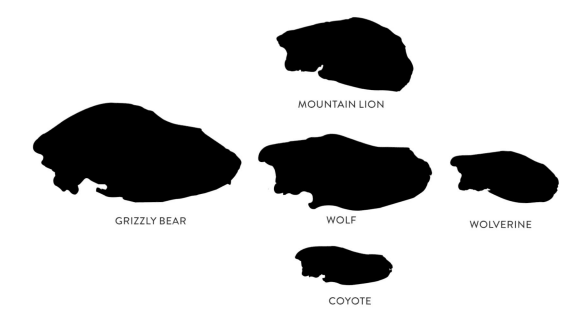

MOUNTAIN LION

GRIZZLY BEAR

WOLF

WOLVERINE

COYOTE

The auditory bullae, located at the base of the skull, house the hearing apparatus. On wolves they are moderate in size, about average for members of the order Carnivora. Wolves, like dogs, can hear higher pitches than humans can, but otherwise the sensitivity of their hearing is similar to that of humans. Their erect and mobile ears likely increase their ability to funnel faint sounds into their eardrums and determine the direction from which a sound originates more accurately than is typical in humans.

The slender rostrum and reduced sagital crest of a coyote fit this animal's omnivorous lifestyle. The blockier rostrum and larger sagital crest of the wolf gives its skull more strength for capturing prey and enduring the forces applied to it in the process of subduing large animals. The mountain lion skull is even more specialized for killing and consuming large prey. The reduced rostrum and mandible gives greater leverage to their canines. Mountain lions and other felids have a reduced number of teeth, all designed for killing or consuming prey, which allows for the shorter skull structure.

The sagittal crest is where the muscles that move the mandible attach. Its large size on a wolf accommodates well-developed muscles that give wolves exceptional bite strength. The temporomandibular joint, where the mandible (lower jaw) hinges on the skull, is well developed and allows for increased stability and strength in the jaw, required for wolves given their propensity to grab hold of large fast-moving objects such as a running moose. It also allows them the bite strength to crack open large bones with their teeth.

Wolf dentition is relatively unspecialized compared to that of some other carnivores, typical of an animal with a varied and adaptable diet. This lack of

This wolf is rolling in a river-otter latrine. Wolves use their powerful sense of smell in both hunting and foraging as well as for communication with other wolves through scent-marking behavior.

specialization reflects their evolution from a more omnivorous ancestor. However, compared to their evolutionary forbearer and coyotes, wolves have a shortened and broader snout and enlarged canine and carnassial teeth. These features relate to their current highly carnivorous diet, as does their decreased relative neck length which gives them increased leverage in using their mouth and teeth to engage with prey species.

All members of the order Carnivora have two opposing teeth or carnassials (the upper fourth premolar, lower first molar) which have evolved for shearing flesh. They have sharp cutting edges that close in opposition to each other like scissors. In wolves, the carnassials are relatively enlarged with the forward portion of the lower one blade-like for shearing and the back portion flattened for crushing. The molars adjacent to the carnassials are conical, a powerful design that aids in crushing hard food such as bone. Located at the back of the mandible, adjacent to the temporomandibular joint, these teeth have the greatest leverage and force of any in the mouth, and allow wolves to crack open the large bones of their prey to access the nutritious marrow within. Mountain lions and other felines lack these crushing structures in their carnassials and also lack posterior molars, which makes them not as adept at breaking large bones as wolves or wolverines (*Gulo gulo*, the largest member of the family Mustelidae) are. Bears also have large molars, which are used primarily for consuming plant material but can also be used to crush bones.

Wolves use their large fanglike canine teeth, designed to puncture and hold onto fleeing animals, for hunting and killing large prey. These teeth must be able to withstand substantial forces applied to them. Wolves are often dragged or lifted off the ground while locked onto the legs or nose of prey. Canines are the most commonly broken teeth in all carnivores; they are also broken more often among wolves that prey on moose as opposed to smaller and less aggressive elk and deer. Canines are worn down over the lifespan of a wild wolf, which is why their length is used to estimate the age of sedated or dead wolves in the field.

Incisors, the small sharp teeth in the front of the skull and mandible, are used for grasping and slashing prey. The outward curved arrangement of these teeth also allows them to be used independently of the canine teeth for such things as removing meat from bones.

LEGS AND LOCOMOTION:
HUNTING AND TRAVEL BEHAVIOR

Searching for wolves, whether across the rugged rainforests of the Pacific Coast or the steep ridges of the Cascades, one thing you learn quickly is that they cover a lot of ground. You may find tracks at first light made the night before, but that wolf might already be twenty miles away.

The foot structure of all members of the family Canidae is digitigrade—when they move, they place only their toes on the ground—an evolutionary adaptation which increases running speed and efficiency. Wolves have extended their limb structure by carrying their weight on their toes and metacarpal bones while preserving the ability of their feet to carry out other tasks as well, such as digging and slashing prey. The rear portion of each foot does not touch the ground, adding length to the limb and increasing the potential stride of the ani-

mal. By contrast bears, who are not particularly efficient runners, carry their weight on their entire foot.

Interestingly wolves' primary prey, hoofed mammals, have evolved an even further specialized foot structure described as unguligrade (hence the term *ungulates*). In an unguligrade foot, the animal lands on the tips of the digits, allowing the toe bones to be added to the length of the limb. This further increases the potential stride and speed of the animal. In general, ungulates' top speeds are greater than wolves', giving them a slight edge. Wolves must make up for this by hunting weakened animals, using surprise, or using terrain that slows down their prey.

Why is it that canids did not also evolve to an unguligrade posture? The answer is in the multiple ways in which they use their feet. Ungulate appendages are highly simplified, designed basically to absorb shock from locomotion and allow for traction in commonly encountered conditions. Wolves and other digitigrade carnivores must also use their feet and claws for catching or manipulating prey. In addition, wolves use their feet and claws for digging, which they do to excavate burrows for natal dens, for foraging for small mammals, and for caching excess food from kills.

Wolf digs, such as this one in north-central Washington, are typically tapered toward the deepest portion of the excavation with a throw mound adjacent to the hole. In my experience, with the exception of dens, wolf digs are rarely more than a foot deep.

Two sibling wolves play on a remote beach, showing off their running prowess. Clayoquot Sound, British Columbia.

The claws of wolves are relatively broad, more blunt, and less curved than those of felines and mustelids, making them relatively useless for climbing trees. Additionally, over evolutionary time, canids have lost one digit from their hind feet, and the inside digit on the front foot is greatly reduced and vestigial, while they have also lost their ability to rotate their front feet. This structural simplification of the feet reduces their functionality for grasping. In comparison, mustelids still retain all five digits on both front and hind limbs, and felines retain five functional toes on their front feet as well as the ability to rotate their forepaws.

Both these families of carnivores can use their feet more effectively for grasping and holding onto prey than wolves can. Because of this challenge, wolves usually kill large prey in a dramatically different way than mountain lions do. Wolves rely on many bites opportunistically placed on the prey, rather than attempting to hold onto it and deliver a single killing bite, as is typical for mountain lions. Some believe that this less efficient hunting method is part of the reason wolves have evolved as social hunters. Conversely, a wolf's toe and claw structure adds to efficiency in running and digging, two attributes that are important elements in a wolf's livelihood. Indeed, wolves (and other members of the canid family) are far more efficient runners than any of our region's other carnivores.

HALF LIFE SIZE

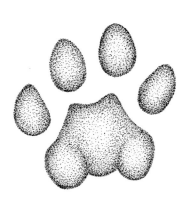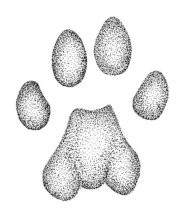

left front   left hind

MOUNTAIN LION

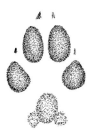

left front   left hind

COYOTE

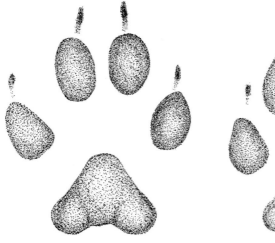
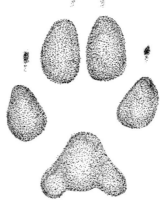

left front   left hind

GRAY WOLF

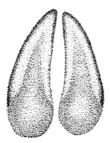
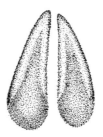

left front   left hind

MULE DEER

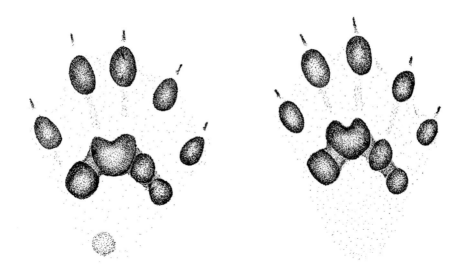

left front   left hind

WOLVERINE

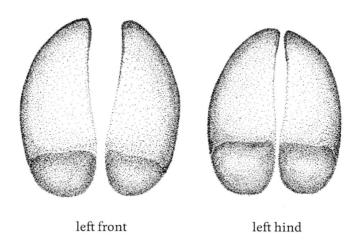

left front   left hind

ELK

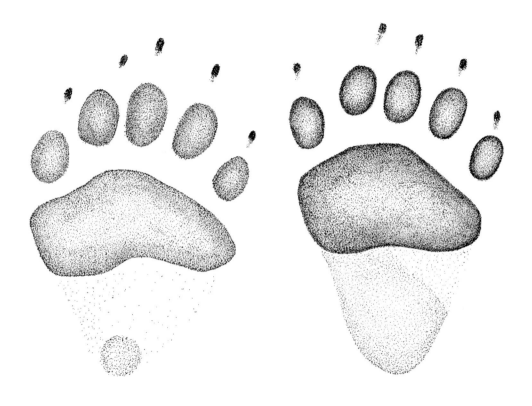

left front   left hind

BLACK BEAR

DIGESTIVE SYSTEM
AND DIET

As anyone who has a dog knows, they can eat just about anything and if it doesn't agree with them, it will come back out one end or the other fairly quickly. Wolves, like dogs, have a simple digestive system designed to accommodate large quantities of relatively easily digestible animal tissue as well as just about anything else that they discover and decide might be edible. In comparison, herbivores such as hoofed mammals, which eat large quantities of hard-to-digest plant material, require much more intricate digestive systems that allow them to break down and process their very specific food. The flexibility of their digestive process allows wolves to adapt to a variety of prey sources and environmental conditions.

Wolves are also particularly well adapted to a lifestyle that involves sudden abundances of food followed by days of scarcity. They can consume up to thirty pounds of meat at a time, but they can also go without food for long periods, an important capability for an animal with highly variable feeding opportunities. When food is scarce they metabolize their own stored body fat and then their marrow fat before their body begins to consume their own muscle. Following such a fast, they can recover weight in just a few days once food is available. In typical conditions in the wild, wolves probably live close to the edge nutritionally, usually carrying a relatively small amount of body fat. Aside from allowing them to regain weight, the ability to consume large quantities of food quickly also helps a social carnivore compete with pack mates for food at a carcass. Wolves sometimes remove parts of a carcass and cache them in a hidden location for later consumption. They can also consume and then regurgitate large quantities of food, which they will do either into a cache or for pups at a den or rendezvous site.

REPRODUCTION AND
PUP REARING

While a wolf pack often has more than one reproductively mature female, only a single litter tends to be produced in a year. Social or nutritional stress may prevent subordinate females from coming into estrus, or they may breed and then reabsorb the fetus because of nutritional stress. In some instances where enough food is available, multiple litters occur in a pack and both litters are reared, either separately or together. In expanding populations the chances of having more than one breeding female in a pack appears to be increased, perhaps because the breeding female has a higher tolerance of daughters breeding within the current pack structure at times of decreased food stress.

The youngest wolf documented to have given birth in the wild was two years old, but in general females won't have the chance to breed until they reach four years. In most instances, for a young female to breed she will have to establish a territory

The swollen teats of this classically colored gray wolf indicates this is a lactating, breeding female. Northwestern Montana.

with sufficient resources to provide nutrition for her and her young. In an expanding or persecuted population (as is the case across much of the Pacific Northwest), young of both sexes may breed relatively sooner than in landscapes with a more stable wolf population. This earlier breeding can occur when young wolves have an opportunity to establish a new territory or when there are enough resources to support more than one litter within a given pack. Older females may be replaced by their daughter as the breeder in a pack but still stay in the pack. Wolves as old as eleven have been documented to rear a litter in the wild successfully.

Female wolves have a single period of estrus annually, its onset linked to expanding daylight hours in the second half of winter. Because of this, as you go farther north the timing of various parts of the breeding cycle shifts to later in the year, or earlier as you go south. Male wolves are also only reproductive for a brief period of the year, also related to photoperiod (luckily for wolves, at the same time as females). At our latitude, breeding usually occurs in late February. For about two weeks prior to estrus and breeding, driven by hormonal changes, the female's attractiveness to males increases and courtship activities, such as play behavior and traveling together, occur—but the female does not allow the male to mount her. A small amount of blood may be found in her urine at this time, a sign that can often be observed in locations with snow on the ground. Estrus, when the female is receptive to breeding, lasts for about nine days. During estrus, the female exhibits positive sexual behavior toward the male and allows mounting and copulation. Gestation takes about sixty-three days with pups—five or six in a litter, on average—typically born in the end of April here in the Pacific Northwest.

Pups are usually born in a natal den dug by the female, or occasionally in a natural shelter such as a hollow tree or under a fallen log. Wolf pups are born blind and deaf. Their eyes open at ten to fourteen days of age, and they are weaned from milk at five to nine weeks as they begin to consume meat regurgitated by adults. By mid July they are ready to move to a rendezvous site up to several miles from the den site. However, some coastal packs, with den sites associated with salmon-bearing streams, will continue to utilize the area around the den site through the summer months. Pups grow at a fast rate for the first six to seven months at which time their adult dentition comes in completely. After this, growth slows considerably but continues until twelve to fourteen months of age. After its first year of life, a wolf will continue to put on mass in appropriate conditions. By the fall of their first year they are ready to travel with the rest of the pack.

Wolves have relatively high reproductive potential for carnivores their size, and expanding and persecuted wolf populations often demonstrate an increased reproductive rate compared to stable or saturated populations. This productivity allows wolves to compensate for losses of adult members and to increase population opportunistically and quickly. It also increases their potential impacts on their environment, allowing them to take advantage of favorable conditions. In quality habitat with low wolf density, food resources are generally high both in terms of abundance and vulnerability. This increases access to nutrition for wolves and decreases social stress around food resources. Increased access to food can increase the chance of pup survival, and indeed pups in expanding and exploited populations have a higher survival rate.

Their high reproductive rate also makes wolf populations fairly resilient in the face of moderate levels of persecution, far more than many other large carnivores, such as bears, which have much lower reproductive rates. In grizzly bears, females give birth to a single litter of one to three cubs every two or, more typically, three years. In a single four-year period, a grizzly sow would not have more than six offspring under the most ideal conditions, and more likely as few as two. During that same time a wolf might whelp twenty-four pups.

LIFE SPAN

The life expectancy of a wolf in the wild in the Pacific Northwest varies widely, primarily depending on the level of human persecution in the area. Wolf researcher and conservationist Chris Darimont reflected on a conversation with David Person, another biologist. Person, who studied heavily hunted and trapped wolves in southeastern Alaska, looked at photos of wolves from Darimont's study area, along the remote central coast of British Columbia, and noted their gray muzzles. He stated that you just don't see wolves with gray muzzles (a sign of age, as in dogs) in his study area where the average life expectancy is about four years. In more stable wild populations, if wolves survive their first year of life, they typically live to the age of ten, exceptionally to the age of sixteen.

Regardless of where they live, wolves in the wild rarely die of old age. In most populations, humans are the largest source of mortality for adult wolves. Even in legally protected populations, poaching and official wolf-control activities can be amazingly high. Between 1979 and 2002 in Wisconsin, when wolves there were legally protected under the Federal Endangered Species Act, poaching accounted for 26 to 72 percent of all mortality for wolves which amounted to potentially more than one quarter of the entire population annually. Besides direct, intentional killing of wolves, accidental deaths from motor vehicle collisions are another source of human-caused mortality.

Besides humans, the most likely animal to kill an adult wolf is another wolf. Disputes over territory with adjacent packs, aggression toward transient animals found within a pack's territory, and occasionally power struggles within a pack all can result in fatalities within a wolf community. Grizzly bears and cougars have been documented to kill wolves occasionally (and vice versa). Hoofed mammals can also kill or severely wound wolves in self-defense during a hunt, often by kicking the wolf. Skull and jaw fractures are two common injuries from such encounters. In times of food shortage, starvation can also take its toll on a wolf population, with pups being most at risk.

Disease can be another source of mortality in wolves. The most pressing disease for wolves in our region is canine parvovirus, a virus that causes severe diarrhea which can lead to dehydration and ultimately death. Parvovirus was introduced into wild canids in North America from domestic dogs in the 1970s and is now endemic across the continent. Parvovirus outbreaks in the neighboring Rockies have had significant impacts on pup survival in some years and received a lot of media attention during the recovery of wolves in that region. The cause of outbreaks is not known specifically, although malnutrition or other conditions that weaken pups could be contributing factors.

LIFE
IN A PACK

Essentially an extended family, the wolf pack is the fundamental unit of a wolf population. While lone wolves make up an average of 10 to 15 percent of any given population, these are mostly dispersing animals seeking to establish their own breeding territory and pack. The typical pack structure consists of an adult breeding pair and their offspring, including pups and possibly the young from previous years. Occasionally other adult animals, either unrelated or siblings of the breeding pair, also join a pack. In persecuted or expanding populations, the percentage of a pack made up by pups is generally higher, either because of increased adult mortality or because two- to three-year-old wolves disperse more easily to new territory. If one of the breeders in a pack is killed, the remaining adult will often maintain the territory and attempt to find a new mate. Occasionally such a pack will disperse, individuals becoming floaters until they find a new mate and territory or join another pack, potentially usurping a breeder there. If one or both of the breeders in a pack die, a young animal from the pack may usurp control of the territory.

The breeding animals in a pack, generally referred to as the alpha male and female, form a consistent pair bond, often for life. The alpha male is usually the most aggressive in hunting and asserts dominance over all other pack members. The alpha female is the most dominant female and asserts control over everything related to pup rearing including selecting den location and digging the den. The alpha female is often the most aggressive in defending pups and den site from transgressors such as bears and wolves from outside the pack.

After the pups are weaned, an animal other than the alpha female will often stay with the pups. This babysitter wolf is usually an older sibling of the pups. In one pack on the British Columbia coast, researchers documented this role being filled for several years within one pack by a wolf with a significant leg injury that reduced its ability to travel and hunt. In other instances a variety of adult wolves within a pack can fill this role in any given season. Once the pups are larger they will be left alone while all pack members participate in hunting forays.

The number of animals in a particular pack is fluid, changing with the departure of young wolves, mortality, and the sudden influx of pups in the spring. The destruction of Washington's Lookout pack vividly demonstrated this fluidity. Studies of wolves from around North America show variations in average pack size

from three animals up to eleven. Rarely have packs with more than twenty animals been reported.

A number of studies have shown a link between the size of packs and the size of the primary prey species. For instance, packs that prey primarily on moose would be expected to be larger than those that focus on deer. However, this correlation is far from uniform as packs that prey primarily on elk, as in central Idaho, and possibly the Imnaha pack of northeastern Oregon, average larger than packs that focus on either moose or deer. Furthermore, packs exposed to a high level of human-caused mortality, such as the Lookout

A single adult wolf with three excited pups milling about is likely their mother. Central coast, British Columbia.

pack, have decreased pack sizes regardless of their primary prey. Reductions in food availability decrease reproductive success and increase dispersal of offspring, but numerous studies have shown that overall prey population densities apparently do not strongly influence pack size in general.

At two to three years of age, young wolves often leave their natal pack to seek out an opportunity to breed. This dispersal may be linked to nutritional stress as the alpha pair limits access to kills if prey is scarce. During this very dangerous time in their lives, these lone wolves may travel around the periphery of their natal pack's territory or disperse hundreds of miles. Generally wolves from a pack act aggressively toward transient animals in their territory. It is not uncommon for boundary transgressions to end in the death of the transient wolf. Besides aggression with other wolves, dispersing wolves face increased exposure to hazards such as road crossings and generally negative interactions with humans.

The most typical way for a new pack to form is for a dispersing male and female from separate packs to find each other, carve out a territory, and breed. This territory might be a landscape previously unoccupied by any wolves. Alternately, the new breeding pair could take up residence in the space between territories of existing packs, or they could carve out part of one of their natal territories. During a range expansion such as what is occurring in the Pacific Northwest, a new breeding pair may occupy a territory adjacent to an existing pack's or up to several hundred miles from the closest adjacent pack. In these instances the male and female may travel together or find each other after independent dispersals from very different places. During such long-distance dispersals, wolves may pass through many areas with excellent prey populations with or without wolves before selecting a place to establish a territory.

Territoriality may be the single most important factor in the distribution of wolves in landscapes they inhabit. A wolf pack's home range (all the land that it uses in a year) must be large enough for adults to find enough food to ensure the survival of pups. The term *territory* refers specifically to area defended against other members of the same species. Wolf packs in the Pacific Northwest actively and vigorously defend their home range from incursions by other wolves. Wolves regularly patrol their territory and maintain boundaries through scent marking, howling, and actual aggression toward other wolves. Scent marking and aggression toward outside wolves is most common during the breeding and pup-rearing season. There may be a buffer zone between the territories of adjacent packs, an area where both packs spend little time. These buffer zones may become a refuge for prey species. Coyotes also make use of these zones in order to avoid altercations with wolves. Territorial boundaries may also consist of low-quality habitat or a large body of water rather than an adjacent pack's territory. In expanding populations, a pack will define its territory even though there may not be an adjacent pack to defend it from.

Territory size can vary widely depending on prey density and overall wolf population density. In expanding populations, territory size might be larger than when saturation of high-quality habitat forces more packs into a smaller space. Home range size can also be linked to prey biomass, with more productive landscapes supporting wolves on smaller territories. Here in the Pacific Northwest, home range sizes have been documented from a mere twenty-two square miles on Vancouver Island to over 850 square miles in northeastern Oregon. Average home range sizes in northwestern Montana and Idaho are about 190 square miles and 225 square miles, respectively, while on the central coast of British Columbia, home ranges average smaller, about 112 square miles.

In some parts of the Pacific Northwest, wolf packs move to higher elevations during the late summer and fall, following migratory ungulates. This was the case with the Lookout pack and may be true of other packs in the North Cascades. Still others, such as the Diamond pack in the Selkirks, continue to use the same areas throughout the year, probably depending primarily on nonmigratory moose in the winter.

THE WOLF'S YEAR

Wolf packs in the Pacific Northwest go through a relatively repetitive cycle of activity over the course of a year, revolving around pup rearing and, in some instances, the migration of prey species, salmon runs, or availability of other locally abundant food. Springtime focuses on preparing for denning activities and ends with the birth and early development of pups. Packs that move up and down in elevation during the year will den at lower elevations where spring comes early and migratory ungulates linger. Hunting activity may focus on newborn elk, moose, and deer, a relatively easy food source. Travel for adult wolves is generally limited to hunting forays within a single day of the den site.

As spring rolls into summer, pups are weaned and begin eating meat brought back to them by adults, but are still limited in mobility. Pups will often be moved away from the den to a new location referred to as a rendezvous site. This move may be prompted by the desire to bring them closer to productive hunting areas or a recently killed animal. Depletion of hunting resources in the immediate vicinity of the den, fleas, and other parasites might also drive wolves to move pups away from the den location. However, in coastal packs, wolves often linger around the den site through the summer into the fall when salmon runs peak. By late summer adults will make longer forays away from the rendezvous site, eventually accompanied by pups.

By fall pups are able to travel with adults, and the pack becomes very mobile. Many packs wander far and wide across their home range, making use of areas that were out of range when the pups were less mobile. On the coast, packs may remain in the vicinity of pup-rearing activities, taking advantage of the fall salmon runs, generally the largest of the year.

With the onset of winter, activity will shift based on the movement patterns of prey species. Wolves may follow deer and elk down in elevation to winter range or shift from extensive use of wetlands to timbered slopes and clearcuts following shifts in areas frequented by moose. On the coast, old growth forests become increasingly important to deer and elk—and wolves. In places with long hard winters, hoofed mammals weakened or killed by malnutrition become a focus of foraging activities. Courtship and territorial activities ramp up in midwinter leading into breeding in the end of February.

THE SOCIAL HUNTERS

Why do canid species become more carnivorous and more social as they become larger? As we have seen, wolves' unique body structure, a remnant of their evolutionary ancestry as omnivores, is part of the answer. While the increased size and compactness of their skull and teeth, as well as their reduced neck length, show evolutionary movement toward greater efficiency in killing large game, wolves have to deal with the body they have now, not the one they might get in another million years as this process continues. While granting them the tools for a unique and powerful means of hunting, their big noses, long snouts, and efficient-for-travel but less dexterous feet come at the cost of efficiency in actually killing large game. There is another example in the evolutionary record of a species that made a relatively late switch over to hunting large game, using social hunting tactics to make up for a lack of body structures suitable for killing large animals—a primate, in fact. One can only imagine it was comforting to have several compatriots at one's side out on the tundra while trying to subdue a mammoth with wooden spears.

But why the focus on meat, and specifically on hunting large game? Here wolves are not unique in the world of large carnivores. Across the globe, once over a certain size (about forty-seven pounds), carnivores are far more likely to be exclusively carnivorous and to focus on large prey. This tendency is thought to be about efficiency in meeting the increased metabolic needs associated with large body size. If you're a large carnivore you will need lots of meat so killing large animals makes a lot of sense. Without a focus on large prey, big carnivores must make other major adjustments to survive. Black and grizzly bears are perfect examples. Both species are highly omnivorous and in many instances get more calories from plant foods than from animals. Plant tissue is generally much lower in energy than animal tissue. One way they make up for the inefficiency in their energy source is to be miserly about energy expenditures—they sleep several months of the year. When they are awake all they want to do is eat, and although they can get by on berries and grass, they go far out of their way to access large prey whenever possible. Grizzly bears often co-opt wolf and mountain lion kills, and black bears will often take mountain lion kills. And here we find another potential driver of wolves toward sociality—efficiency in consuming the large prey they have come to depend on. A single wolf or pair loses substantially more food to scavengers from a kill than does a larger pack.

Wolves are scent-driven hunters designed to cover vast stretches of ground. They hunt primarily by chasing rather than by ambushing prey. Their social nature increases their resources for hunting, while their metabolic yearning requires satiation in the form of meat on a large scale. Their social structure also allows them to increase their reproductive rate quickly when conditions allow, as many packs have more than one reproductive female available if food is abundant. These elements, combined with the propensity for some young wolves to disperse hundreds of miles from their birthplace, have created a powerful and dynamic ecological force. Driven by a biological urge to reproduce and defined by unique bodies and behaviors developed through eons of evolution, wolves push hard to establish a home, build a family, and feed their young, efforts that can send ecological ripples through the landscape. Once again, our own species comes to mind.

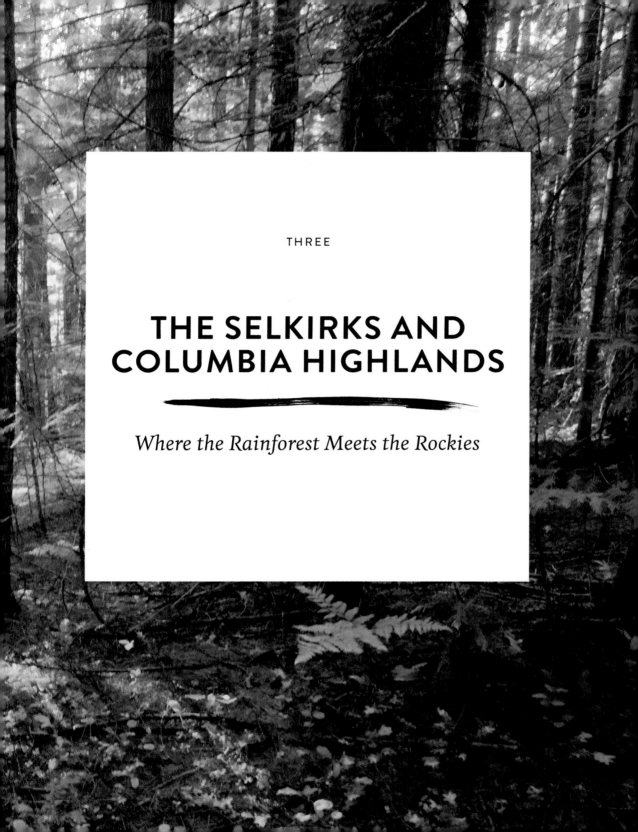

THREE

THE SELKIRKS AND COLUMBIA HIGHLANDS

Where the Rainforest Meets the Rockies

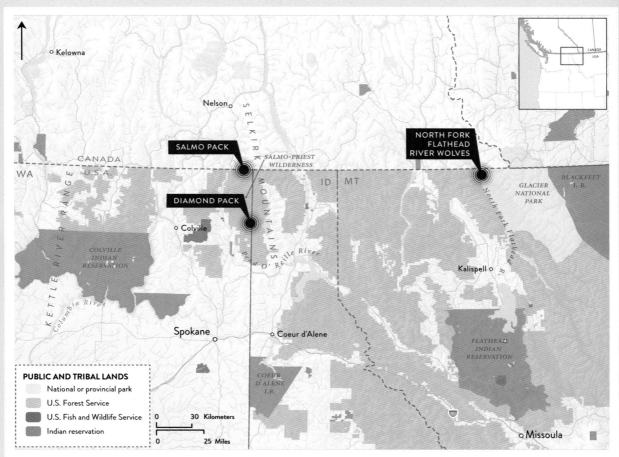

COLUMBIA HIGHLANDS, SELKIRK MOUNTAINS,
AND NORTHERN ROCKY MOUNTAINS

**B**ELATEDLY, I REALIZED MY FOLLY. The route I had chosen seemed like a good one on the map. "Typical," I thought to myself as I clawed my way on steep slopes through thickets of huckleberry and mountain ash, dense stands of conifers, and mucky-bottomed slide alder chutes, attempting to reach a ridgeline beyond which I hoped gentler terrain awaited.

Earlier as I picked my way along a game trail through thick subalpine forest, the grunt and crashing of a bear close by made my heart skip a beat as I instinctively reached for my camera and then realized that perhaps my bear spray would have been more appropriate. I had quickly moved out into the relative security of a small wet meadow where I could see more of what was around me, only to discover fresh grizzly bear tracks filling up with water. With this experience fresh in my mind, I was grateful to discover a slight trail weaving away from the worst brush, under a cliff band and toward the pass on the ridge that was my target.

Reaching the pass I expectantly gazed over the other side and breathed a sigh of relief at the open slopes leading down to a small tarn. Several miles and a couple thousand feet beyond that, Kootenay Pass, Canadian Highway 3, and my truck waited, parked just a few miles as the crow flies from the Washington border. Relaxing, I sat down, only then looking over to see a long tubular scat filled with ungulate hair and bone fragments. Wolf? Up here? Where am I?

In a daylong tour of the mountains along the Washington, Idaho, and British Columbia border, I felt that perhaps I had been magically transported to the Chugach Mountains of Alaska. Signs of caribou, moose, and grizzly bear greeted me all along my trail. Reaching this small pass and finding wolf sign really shouldn't have been surprising in this context. With mountains rolling away in all directions, a sense of wildness came over me very different from what I'd experienced elsewhere in the Pacific Northwest much farther from a highway but without these megafauna occupants.

The Selkirks are a bit of a forgotten mountain range, lying on the western edge of the Rocky Mountains and tucked between this continent-spanning range and the rain-soaked and glacier-clad Cascades and British Columbia Coast Range to the west. Besides being forgotten by the humans of our region, the Selkirks also seem

---

PREVIOUS: High peaks of the Selkirk Mountains in the Salmo-Priest Wilderness along the Idaho-Washington border.

a bit confused ecologically speaking. Here dense forests of western hemlock and western red-cedar (denizens of the rainforests west of the Cascades) mingle with lodgepole pines, Engelmann spruce, and subalpine firs (staples of colder, drier interior mountain ranges), with an additional smattering of ponderosa pines (a drought-tolerant pine typical of warm dry slopes). Birds as well as mammals large and small typical of landscapes much farther north find the southern edge of their range in forests of the Selkirk Mountains. Boreal chickadees and northern hawk owls haunt dense forest while northern bog lemmings make their home in among the wet meadow grasses and wildflowers, and Canada lynx hunt snowshoe hares in the spruce forests.

The Selkirks are home to a rare plant community: inland temperate rainforest. Temperate rainforests of any sort are unusual across the globe (representing about one one-thousandth of the earth's land surface), and the Pacific Northwest is home to half of all such rainforests. But conditions that produce over one hundred inches of precipitation inland in temperate regions of the world are even more limited. Patches of rainforest appear in deep mountain canyons where looming stands of red-cedars line the banks of creeks, which until recently, despite being hundreds of river miles from the ocean, were the destination of robust salmon runs. Unlike the Cascades to the west, here the rainforests host herds of caribou, which spend part of every winter in the shelter of the forest, escaping the deep, unconsolidated early winter snowpack of the higher elevations of the mountains.

These beautiful forests have not, however, been overlooked by the timber industry. Exploring most of the range now you are more likely to find big stumps than big trees amidst a flush of younger stands of trees. Similarly, the rivers of the region have been a prime target for hydroelectric power. The long fingers of Seattle reach all the way out here, and the salmon runs of the past have given way to summer lake fishing for locals and street lights and refrigerators for the Northwest's largest metropolis.

In this landscape of ecological paradoxes, pockets of rainforest and moist, high-elevation subalpine and alpine zones contrast with arid timberland that are subject to relatively frequent natural fires. In these stands, the area's historic fire regimes maintained a patchwork of open parklike pine forests, ponderosa pines in some, revolving stands of thick lodgepole pine in others, and aspen forests in slightly moister landscapes where fire burned off conifer stands. On large stretches of these forests in the Selkirks and across the Columbia Highlands (the broader mountainous region that connects the Rockies to the Cascades), suppression of naturally occurring fires has left another mark on the landscape. Dense stands of Douglas fir crowd around the stately old ponderosa pines in some areas. In others, dog-hair thickets of lodgepole pines grow so thick that travel through them is nearly impossible. Such crowded tree stands create stressful growing conditions for individual trees, leaving the entire forest more susceptible to outbreaks of beetle infestations. Crowding also leads to conditions ripe for very large stand-replacing fires, something less common in the past when more of the forest comprised larger, more widely spaced trees lacking lower limbs. In those conditions, ground fires often burned brush and fallen limbs while mature trees survived. Even when old trees remain on the landscape, the flush of younger trees crowding around them can act like a ladder, carrying fire up past the limbless and thickly barked lower portions of the tree into its crown and killing it.

Home to two large Hudson's Bay Company trading posts during the 1800s, the region saw extensive trafficking in wolf pelts, but by the second half of the century the animal had become scarce. From the late 1800s until the 1990s, there was only sporadic documentation of wolves in this area. These animals were either remnants of the original population or dispersers from farther north and east in British Columbia and Alberta. In the 1970s wolves began reestablishing themselves in the southern portion of the Rockies in Canada and eventually south across the border into Montana. With this population established to the east, wolf dispersals increased into the Idaho panhandle and the Selkirks. By the start of the twenty-first century, wolves were once again clearly reestablishing themselves here. The genetics of wolves in the Selkirks and Columbia Highlands link them to wolves from the Rockies in northwestern Montana and southern British Columbia and Alberta, according to WDFW data. In 2010 and 2011, I went in search of signs of the first two packs discovered in the remote mountains and dense forests in the northeastern corner of Washington State.

IN THE
DIAMOND PACK'S TERRITORY

The howling had been going on since before first light when I had arrived in my blind. Located on the edge of an old logging landing, the blind was about a half mile north of the den of Washington's Diamond pack. The 4 a.m. spring twilight had grown to the full light of day, and the lyrical songs of Swainson's thrushes had faded as the day warmed. But the howling persisted. Now, seven hours later, the howling was clearly originating from two distinct locations. From my best efforts to plot the direction of the noise on my topographic map, I gauged that it was originating from the den site and then to the west of the den, close to a large wetland. What was going on? Perhaps the pack had made a kill down in the wetland and communication was going back and forth between that location and wolves at the den with pups. Finally, after a series of howls from several wolves down by the wetland, a resounding response echoed from the den, starting with the low moan of an adult and then erupting into a cacophony of higher pitched yelps and barks of pups.

When discovered in 2009 the Diamond pack became Washington's second confirmed pack. Its territory spans the Washington-Idaho border in the Selkirk Mountains and is similar in size to that of the Lookout pack, about 383 square miles

according to the WDFW. About two years later, for nearly two weeks, I had been observing the pack through their tracks, signs, and howling. I had spent between eight and twelve hours most of those days in a blind along a commonly used travel route of theirs and within earshot of their howling from the den. Over the course of these observations I pieced together a great deal about the pack, including diet information, habitat use, travel routes, the general location of their den, and that they had pups again this spring. But I had yet to actually lay eyes on one of them.

The repetitive howling coming from the den site and the wetland made it pretty clear that, first, I knew where much of the pack was at the moment, and second, they didn't appear to be planning on swinging by my blind anytime soon. Having no interest in disturbing the den site at this critical time of the year, I decided to try to sneak into the wetland to see if I could learn what they were howling about and possibly get a glimpse of an animal in the pack.

Stalking up on a wild animal, especially one as aware as a wolf, always involves a certain amount of luck if you're to succeed in observing the creature without being detected yourself. Even if you plan and execute everything perfectly, things that are beyond your control can reveal you to your quarry: the wind can swirl and carry your scent to them, they can happen to look in your direction at just the wrong moment, the substrate you must pass over could be impossibly noisy, or the animal may move on before you can approach closely. Wolves themselves fail more often than not in their attempts to apprehend their prey. That we humans are also imperfect in this regard is really only natural and makes success that much sweeter. In the ten years since I first encountered a wolf in the wild, I had never stalked and observed a wolf undetected. With this in mind I grabbed my camera, snuck out of my blind, and headed for the wetland.

This wet meadow system close to the Idaho-Washington border in the Selkirk Mountains has often been frequented by the Diamond pack.

Two hours later found me exhausted and exhilarated, standing in water up to my knees in a flooded patch of alders in the middle of an old beaver pond, staring through my camera lens at a young wolf as it nervously trotted through the brush on the edge of the pond. It was not nervous about me, however. It looked over its shoulder in the direction most of the howling had been coming from, clearly attentive to the rest of its pack which was not too distant. The running water coming into the pond drowned out the sound of my shutter as I attempted to catch a frame of the wolf in the brief moments it fully emerged from the brush.

Over the course of the next ten minutes I watched and heard it nervously scamper back and forth around the wetland, stopping to glance back anxiously to where it had come from and to return howls coming from there as well. Most of the time I could only see its broken outline through the brush, but in the few brief moments when it fully emerged, I could make out its light gray color and the slender appearance of a young wolf that had not yet grown to its full mass. Eventually it disappeared back the way it came, toward the rest of the wolves, and I retreated to my blind.

A large section of the Diamond pack's territory is a "checkerboard" landscape. I first learned about checkerboard landscapes during my eighth grade United States history class. As an incentive to get a transcontinental railroad built, in the 1860s the U.S. government gave railroad companies every other square mile of land within forty miles of the tracks they laid down. Around the same time the federal government also granted thousands of square-mile sections of land to the state of Washington for the purpose of helping build schools, prisons, and other public buildings in the state. Despite a century of land exchanges and purchases, this patchwork of private, state, and federal lands still exists across landscapes throughout the Pacific Northwest, creating something of a management nightmare as well as an interesting experiment in human land use practices. On the checkerboard lands in the Diamond pack's range, it is equally easy to see the divided land ownership boundaries on a USFS map, where each square mile is alternately colored green for public and white for private, as it is from space, where alternating square miles appear green (with forest cover) for public and brown (heavily logged) for private. Even if you're walking around on the ground, these boundaries are easily noted, carved into the landscape in stark, straight edges where dense stands of forest abut private lands where timber has been essentially liquidated.

The Diamond pack's home range lies south of the high peaks in the Salmo-Priest Wilderness. On the western portion of the range, logging roads zigzag across the landscape and the scars from skidder trails going straight up hillsides are visible years after logging operations are complete. The road system is so complicated that getting accurate maps that include all the roads (many of which are not

signed) is almost impossible, and satellite photographs are often required to find roads in the landscape. Road density is often used as a parameter for determining suitable wolf habitat, and the tangle of roads in this area exceeded densities deemed suitable. For this reason much of the Diamond pack's home range, including the area they have denned in for multiple years, was excluded from several models developed to predict potential wolf distribution. Their persistence here is likely because many of these roads are gated.

OPPOSITE: This yearling wolf from the Diamond pack has ear tags from having been captured and marked by WDFW biologists as a pup the year before. I photographed it at the end of a long stalk into a flooded wetland after hearing howling in the vicinity.

ABOVE: A cow moose traverses a beaver pond in the Selkirk Mountains.

Besides wolves, another species has also been on the increase in the area. Moose populations in the southern Selkirks in Washington and Idaho were reported in very low numbers through the 1800s. The species was protected from hunting in Idaho around the turn of the twentieth century, and population slowly increased in the early 1900s. Drawn to the increased shrub cover in heavily logged areas and protected by carefully managed hunting seasons, moose (*Alces alces*) have been increasing in the southern Selkirks since the 1950s, with a pronounced upturn in the rate of growth since the 1980s. In the winter, the Diamond pack continues to roam the higher elevation forests along the Selkirk Crest, with occasional forays to lower elevations on the east side of the range. During this time of year moose (and the occasional bull elk) are the only hoofed mammals sticking around higher elevations. Most elk and deer migrate down to lower elevations along either flank of the range. While I was searching for the Diamond pack during the winter of 2010–2011, the trails of moose were the most common large animal tracks I encountered. Skiing along snowed-in and gated logging roads, I followed their deep, ponderous, and widely spaced tracks meandering down the roads as they browsed from willow clump to willow clump. Individual tracks were often three feet deep in the snow.

One afternoon, WDFW biologist Jay Shepherd and I watched a moose browse and travel through the deep snow on a nearby hillside among the hummocks of logging slash and stumps on a recently cut slope.

Surpassed among hoofed mammals only by giraffes for leg-length-to-body-size ratio, the long-legged moose is well adapted to wading through deep snow in winter. Moose are primarily browsers, and their long legs give them access to twigs and buds high up on shrubs and trees. In the winter they seek out upland slopes and forests where they forage on the buds and shoots of woody plants, the only thing readily accessible anyway when a deep snowpack settles over the landscape. A relic of the ice age, they have large bodies and other adaptations for surviving in the cold and snow which mean that they easily overheat in the summer. So during the warm months they gravitate toward cool wetlands and shady stream corridors where they browse riparian vegetation and aquatic plants.

Wolves are also adapted for travel in areas with big snowpacks, but unlike moose, they are built to travel over snow rather than through it. Wolves have relatively large feet for their body size, which gives them more flotation on loose snow. This difference in adaptation can give wolves the advantage when they pursue moose in such conditions. Perhaps this is why moose have such an ornery disposition. Where deer and elk will generally flee, a healthy moose, when confronted by wolves, will usually stand its ground to fend off an attack. This tactic is usually effective as the moose's long legs and hard hooves can be deadly for a wolf. If the moose turns to flee, something sick or injured moose are prone to do, the wolves have the advantage.

Without moose, the core of the Diamond pack's home range would likely be devoid of enough prey for their survival in winter, and indeed the WDFW has identified moose as an important food source for this pack. Farther north, in similar habitat in British Columbia, moose has been identified as the primary prey for wolves. Recent increases and expansions of wolf populations across the interior of British Columbia have been linked to the expansion of moose populations which began growing in the 1930s following logging activity, as has been seen in the Diamond pack's range. A similar pattern has even been documented on the other end of the continent in the eastern boreal forests of Ontario. In the Selkirk Mountains, besides the forests, there is another loser in this equation though: caribou. And of course, wherever a hoofed mammal is involved, wolves may be close behind.

On one of my trips out to the Selkirks to search for wolves I stopped into the Colville National Forest ranger station at Sullivan Lake to chat with wildlife biologist Michael Borysewicz. Walking into his office I noted a weathered-looking antler sitting on top of a set of shelves. The enlarged tine which protrudes forward on the rack, which the animal sometimes uses to shovel snow as it searches for food, identified it clearly as a caribou antler. I asked Borysewicz about it, and he mentioned that a fire crew had picked it up while working in the area.

When most people think of caribou (*Rangifer tarandus*), they probably imagine enormous herds of barren ground caribou (*R. t. groenlandicus*) streaming across the open tundra. However, mountain caribou, a specific subset of woodland caribou (*R. t. caribou*), the subspecies that inhabits the mountains of the interior of our region, travel in smaller groups and frequent forests and steep mountainous terrain. Mountain caribou are another creation of the ice age, well adapted for cold and snow but in a very different manner than moose. Moose, the largest ungulate currently living in the Pacific Northwest, went for big and bold: plowing through the snow, facing down their predators with their aggressive nature and sheer size. Caribou took an approach with a bit more finesse. Relatively slender in build, caribou average smaller in size and weight than elk. Like wolves and other winter-adapted species, caribou have large feet for their body size, for greater flotation when they travel over snow. Their large, round feet are cupped in the middle; in winter they can use them like shovels for pawing through snow to reach plants buried beneath.

Mountain caribou have a unique migratory pattern among our region's mammals. Unlike their northern siblings, which migrate hundreds of miles back and forth across the landscape, mountain caribou migrate up and down the mountainsides, changing altitude rather than latitude. But this is not what makes them unique. They migrate twice each year, including heading up to higher elevations for part of the winter. Mountain caribou spend their summers at high elevations grazing in subalpine and alpine meadows and browsing on subalpine forest shrubs and forbs. With the onset of winter snows, they move down into lower-elevation old growth forests where the snowpack is lighter and they can feed on plants on the forest floor. As winter progresses and snow builds at lower elevations, they migrate back up into subalpine forests where increasing snowpack can act like an elevator, giving the caribou access to their primary winder food source, black tree lichens (*Byoria* species), which they feed on higher up on the trees. With the onset of spring and the melt-off of snow at lower elevations, they migrate back down to take advantage of the early green-up and subsequently follow the melting snow back up to higher elevations, arriving back to high meadows for the summer.

Unlike moose, a caribou's approach to predators is one of avoidance rather than confrontation. Their unusual migratory pattern is the elaborate dance they have developed to stay fed in places where their predators are not likely to be. Deer and elk choose to winter at lower elevations. Moose, being browsers, are not particularly partial to mature forest stands, instead reaching their greatest abundances in locations with more extensive shrub cover. With deer and elk as the primary prey of mountain lions, and moose as well for wolves, these predators tend to associate their winter ranges with these species, leaving the caribou predator-free during the most difficult time of the year.

Mountain caribou are in decline all across the southern portion of their range. In recent years only about forty animals still trek back and forth across the borders between British Columbia, Idaho, and Washington. The decline of caribou populations in these mountains highlights the complex nature of interactions in ecosystems, the adaptable nature of wolves and other large carnivores, and the surprising and unexpected impacts of human consumption on our natural heritage. The plight of mountain caribou was set in motion by industrial logging operations which have removed vital winter habitat, fragmented landscapes, and invited moose into locations where they were previously rare or absent. Wolves, in turn, have responded to this change in moose populations. Deer populations have also increased with a similarly resulting increase in mountain lions in some areas.

Caribou depend on large tracts of old growth forests at both low and high elevations for multiple reasons. At lower elevations, the complex canopy and forest structure of late successional forests reduces the depth of the snowpack on the forest floor and allows caribou to access forest floor plants longer into the winter. In clearcuts the snowpack builds quickly, and in mid-successional forest, forage plants are much less abundant than in older forests. Along with selecting habitat away from other ungulates and large carnivores, caribou further reduce the risk of predation by spreading out across the landscape to reduce their density and therefore detectability. However, the removal of large tracts of old growth forests has also removed wide swaths of habitat and fragmented what remains, forcing caribou into smaller, more concentrated areas, which can both deplete food resources there and make them more vulnerable to predation. At higher elevations, the lichen on which they depend does not start growing in abundance in subalpine forests until tree stands reach at least one hundred years of age. Industrial logging operations, which often go all the way up to treeline, can be so destructive at higher elevations that it takes decades for trees to even begin to establish themselves again, taking these landscapes out of productivity for caribou for well over a century.

Compounding this large-scale loss of habitat is an influx of moose into recently logged areas, which attracts the attention of wolves. This expansion of moose and wolves into new areas undermines the caribou's avoidance strategy. While wolves continue to focus on moose as their primary prey here, it puts them into more contact with caribou who become a secondary prey. Similar changes in the dynamics between deer, mountain lions, and caribou have also been documented in our region. In numerous studies, predation by one or several of these large carnivores has been documented as the primary direct source of mortality for the region's shrinking caribou populations.

This cascade of ecological interactions, which starts with human choices that drastically modify the habitat of caribou, has played into old stories about wolves as out-of-control killers requiring human intervention to "protect" caribou. However, any management strategy that focuses on predator control is ultimately doomed as the problem originates at a much broader ecological level. Without large-scale reform of timber harvest practices and restoration of forest landscapes in the Selkirks and other interior mountain ranges in British Columbia, the factors that have reduced caribou habitat and increased predation pressure will only become more severe. Recognizing this, in 2007 British Columbia's Ministry of Agriculture and Lands recommended, along with the immediate removal of wolves from the vicinity of endangered caribou herds, the gradual reduction of moose populations in these areas through public hunting, the reduction of winter recreational activities which stress caribou, and the cessation of timber harvest activities in their habitat. The ministry declared the protection of 2.2 million hectares (about 8500 square miles) of forest land to aid in the recovery of dwindling mountain caribou populations.

Implementation of this plan has met with mixed reviews. Mineral extraction inside the recovery zone continues, as does mineral exploration, which has been associated with detrimental impacts on caribou. While limited in scope, heli-ski operations and snowmobile use continue in the recovery zone. Timber harvest activities have halted, but much of the recovery zone consists of roaded and logged landscapes for which no restoration activities have been carried out. A more liberal hunting season for moose has reduced their populations in some areas, and various predator control activities have been put into place. These include increases in the length of the hunting season for wolves and cougars as well as an increased limit to the number of animals a hunter can kill. The trapping season for wolves has also been increased, and the province has provided training and traps to trappers as well as offering a bounty for wolves taken from areas with endangered caribou in them. Experiments with sterilization of wolves in areas with endangered caribou have not shown increases in caribou numbers. Plans for shooting wolves from aircraft are in the works. In neighboring Alberta, along with these activities, the province has gone so far as to distribute strychnine-laced bait to kill wolves (and, of course, anything else that consumes it) as part of caribou conservation efforts.

Concerned that loopholes in the recovery plan still existed and that wolves and other carnivores would be killed in the name of caribou recovery (while the root cause of caribou decline—habitat destruction—continues), a coalition of conservation groups has continued to advocate for stronger measures to protect and restore caribou habitat. Indeed, the timber and mining industries are powerful, and with the ease of playing into stereotypes of wolves as wanton killers, it would seem plausible that political pressure would exist to lean heavily on predator control at the expense of the habitat protection and restoration that is the ultimate solution to the current plight of mountain caribou.

In Washington and Idaho, attempts by the USFS to conserve mountain caribou have focused on protecting suitable habitat and efforts to aid in the development of late-successional stand characteristics in areas with younger stands. Timber harvest in designated caribou habitat is limited to activities designed to aid in habitat restoration for caribou, typically through thinning and attempts to favor typical climax forest tree species.

As for wolves, they will keep doing what they do, attempt to find a home where they can raise young and feed a family. Where to from here for the forests, caribou, and wolves? It would appear the answer to that sits in the laps, or perhaps the homes, of us humans. In 2004 I surveyed lumber yards across the western United States to trace the distribution of timber originating from mountain caribou habitat and found that it was carried widely in Washington, Oregon, California, Nevada, New Mexico, and Arizona.

Caribou and humans aren't the only two species with a mixed relationship with wolves in the Selkirks and Columbia Highlands. Grizzly bears, for instance, have a love-hate relationship with their smaller furry neighbor. Watching a grizzly bear feast on the remains of an elk killed by wolves illustrated for me the love part of the relationship.

During the spring of 2008 I was working for biologist Cristina Eisenberg in the valley of the North Fork of the Flathead River in northwestern Montana. Eisenberg was carrying out her doctoral research, studying the relationships between aspen trees, elk, and wolves. Historically, the North Fork is an important location for wolves in the Pacific Northwest— in 1986 it became the site of the first documented breeding pack to return to the northwestern United States, comprising wolves that had dispersed from over the border in the Canadian Rockies. At the time of Eisenberg's research, more than three decades had passed since wolves had returned. She was working to sort out how the return of wolves might have reshaped the ecological landscape in the valley.

After wrapping up a day of fieldwork, we noticed an elk carcass on the ground in a location that we had surveyed earlier that day. As we approached it, a wolf ran off. Knowing that bears have an uncanny ability to find carcasses and a nasty disposition toward anyone else arriving on the scene, we unholstered our bear spray containers and quickly went to assess the carcass. We lingered only long enough to determine what had happened and snap a few photos. The elk had been taken down on the edge of a large meadow system, at the base of a short but steep slope that might have acted as an obstacle for the elk trying to flee or as cover for the approaching wolf or wolves. The elk had just been killed, and as is typical for wolves, the carcass had been opened up in the belly below the rib cage where feeding usually begins. We retreated to our vehicle and left for the evening.

The next morning we returned to continue our fieldwork. Just off the dirt road to the area, we could see a large brown shape hunched over the carcass. As we got closer, the grizzly bear looked up from its prize and glared at us. Even in the safety of a vehicle, it's a little unnerving to have an adult grizzly bear—its face and snout stained with blood, its front feet, with their three-inch-long claws, perched atop the disemboweled remains of what had just recently been an animal three or four times your size—stare at you menacingly. And this is by far the mildest way that a bear will communicate its displeasure with you around such a feast.

We parked the vehicle and watched. After determining that we were not going to approach any closer and attempt to usurp its breakfast, the bear went back to its meal. After about a half hour of feeding, apparently satiated, it tried to drag the carcass off to a more discrete location. But the carcass was a bit too heavy for even the bear to drag up the slope, and after several attempts it abandoned the project and disappeared into the trees. Thankful that our fieldwork was taking us in a slightly different direction, we carried on with our day's work.

That evening the carcass was not visible from the road, and about forty-eight hours after we had initially spotted the dead elk, we returned to the location to see what else had transpired there. All that remained of the elk was a few tufts of hair and its stomach contents. We explored the slope above to try to find more of the carcass but could turn up nothing. We did find tracks, however. Besides the tracks of wolves and the grizzly bear, we also found fresh footprints of black bear, badger, and coyote around where the carcass had been. Apparently the grizzly bear wasn't the only carnivore in the area that was attracted to the wolf-killed elk.

Just as wolves directly and indirectly affect their prey species, so too do they have noticeable impacts on their fellow carnivores. Some of these impacts are fairly straightforward and intuitive, such as an increase in carcasses for scavengers. Others are more subtle, such as the displacement of mountain lions from their own kills, causing the cats to increase their hunting rates. The ecological impacts of these interactions can trickle through ecosystems and ultimately create measureable changes across a landscape.

Wolves are members of the large carnivore guild in the Pacific Northwest, guild referring to a group of species using common resources in a similar way. In our region this guild also includes grizzly bears, black bears, and mountain lions. Across the Pacific Northwest wolves interact with other guild members in different ways depending on the ecological context. For instance, wolves and grizzly bears interact around elk carcasses in interior mountain ranges and at salmon-bearing streams in coastal rainforests. With the extirpation of wolves and grizzly bears from much of the region in the twentieth century, the integrity of this guild and its ecological influence have been significantly altered in many areas. With the return of wolves to the Selkirks and Columbia Highlands and continued recovery of grizzly bears here as well, this guild is once again complete in this part of the Pacific Northwest.

I once heard wildlife biologist Susan Morse refer to bears as a nose and mouth on four legs, alluding to their prodigious appetite and exceptional sense of smell. One thing grizzly bears use their sense of smell to do is to locate the carcasses of animals killed by other carnivores. While not adept at killing adult ungulates themselves, grizzlies are often able to take possession of a carcass wolves have killed.

As might be expected, the relationship between wolves and grizzly bears is often a tenuous one. Most interactions between these two carnivores occur around carcasses. Wolves occasionally kill grizzly bears and vice versa. Wolf diet studies occasionally find grizzly in their diet. Similarly, studies of grizzly bear diets have occasionally documented wolf on the menu. Besides their mutual antagonism, wolves and grizzly bears may together exert broader ecological impacts. Grizzly bears (and black bears) often kill a large number of juvenile ungulates; bear and wolf predation combined has a greater impact on prey species populations than either predator's influence individually. Further, when grizzly bears co-opt carcasses from wolves, wolves appear to increase their kill rates, again amplifying their ecological impact. Wolf-killed carcasses add an important food source for grizzly bears, especially on emergence from their den. This increased food source might increase reproductive rates for grizzly bears, which is of even greater significance here in the Pacific Northwest where they are considered threatened or endangered in various parts of the region.

Black bears, smaller and less aggressive than grizzly bears, usually come out on the losing end of interactions with wolves. Wolves are generally successful at chasing black bears away from carcasses. Others have reported wolves seeking out bears in their den and killing them. In these instances, sometimes the bears are eaten, other times not, hinting that wolves see black bears both as a menu item and as competition.

Like black bears, mountain lions survived the extensive predator control efforts that removed wolves and grizzly bears from much of the Pacific Northwest, though their presence was greatly reduced. Following the cessation of predator control programs, their numbers and range appear to have rebounded. While out searching for the Diamond pack, I encountered their tracks numerous times, including within a quarter mile of the pack's active den location. In the absence of wolves, mountain lions have been the only large carnivore that specializes in killing large prey across much of the region. They are effective solitary predators, capable of killing even bull elk that can be many times their size.

Similar in size and sharing their primary food sources, wolves and mountain lions occasionally have antagonistic interactions. While such altercations can be fatal for either of these competing species, no studies have shown either to be a significant source of mortality for the other. The modern hunting practice of using hounds to tree mountain lions makes use of an evolutionary strategy these large cats have developed to find refuge from wolves, which can't climb trees. Residents living close to the Lookout pack in the North Cascades observed several wolves from this pack tree a mountain lion in the vicinity of a deer carcass. It was unclear who had killed the deer. When wolves (and bears) drive mountain lions off their kills, the kill rate of these large cats has been shown to increase, which in turn can increase the impacts of mountain lions on prey populations.

Mountain lions and wolves typically employ very different hunting strategies. Mountain lions are primarily ambush predators that focus on forested, brushy, and rocky terrain. Wolves, coursing animals, generally prefer to run down prey either in open or forested terrain. The presence of both predators creates a landscape with a mosaic of predation threats that ungulates must carefully navigate. While lingering in open areas away from brush that would be good hiding cover may be a good policy for avoiding mountain lions, it also makes an animal far more likely to be detected by wolves in a landscape that plays to the wolves' propensity to test and chase their prey.

Along the North Fork of the Flathead River in northwest Montana, a landscape similar to much of the Selkirks, winter conditions compress the area used by deer as they seek out lighter snowpack and abundant winter browse plants. Wolves and mountain lions similarly adjust and focus their hunting efforts on these areas. But even hunting in the same locations, these two predators often take different approaches to the task. During a weeklong ski trip in the area, I trailed a mountain lion through a section of forest that was heavily used by mule deer for feeding and bedding. The mountain lion's tracks showed how it slowly and methodically wandered through the area. Several sequences of tracks showed that it stopped, stalked, and then sprung for a brief chase. The explosive tracks of the deer told the other end of the story. In both instances the chases ended in just fifty yards or so, unsuccessfully, with the mountain lion going on its way.

Two days later, I trailed a pair of wolves through the same patch of forest. They approached things very differently. Their long strides showed them moving through the landscape at a trot, a slow run, weaving back and forth across each other's trail. Their tracks cut through areas with heavy deer sign, numerous times literally trotting through the beds of deer in the snow. They also had multiple unsuccessful chases, their evenly spaced trotting tracks bursting into a long galloping pattern and again paired with the explosive fleeing tracks of the deer.

Together, large carnivores exert a powerful ecological force on the landscape. The cumulative effects of their predatory activities can be many times greater than the impacts of individual species for several reasons. Because predators employ different tactics and landscape features for hunting, they can shift both the behavior and patterns of habitat use of prey species who attempt to minimize their risks of being killed. Wolves alone may not be able to significantly alter prey species population size, but in combination with other guild members, their impacts can become significant. Wolves in concert with bears and mountain lions, for example, can have profound impacts on prey, although it can be difficult or impossible to pick out the impacts of one species versus another in these situations.

ABOVE: *A grizzly bear drags away the carcass of an elk which had been killed by wolves the day before. Northwestern Montana.*

OPPOSITE ABOVE: *A grizzly bear walking on the edge of a thick lodge-pole pine stand in northwestern Montana.*

OPPOSITE BELOW: *A black bear emerges from the trees into a subalpine heather field. Trees provide security for black bears which can fall prey to wolves and grizzly bears where their ranges overlap.*

THE ELUSIVE SALMO PACK

Several weeks after my encounters with the Diamond pack, I returned to the Selkirks to search for the Salmo pack, whose home range spans the British Columbia–Washington border, joined variously by friend and colleague Emily Gibson and research assistant Rob Nagel. During the spring the pack appeared to be limiting their travels to the southern portion of their range, indicating that they were likely denning in Washington State. Like the North Cascades, this part of the Selkirks is a federally designated grizzly bear recovery zone. To create conditions more compatible for grizzlies, many of the forest roads are gated and closed, either year round or seasonally. Such road closures reduce conflicts between bears and people and reduce related stresses on bears (and people, I suppose). Because of this, getting access to the Salmo pack in the spring involves a bit more legwork than it might otherwise.

Where the Diamond pack's home range is a rolling landscape defined by large cut blocks and logging roads, the Salmo pack has set up shop in much more striking environs. The pack is named after the Salmo River which traverses the northeastern portion of the Salmo-Priest Wilderness before heading north into Canada. It's a wild and inaccessible landscape. Steep and densely forested ridges rise up for thousands of feet. Even north of the wilderness area, road access is limited. During a week of camping here the only other person we encountered was a border patrol agent watching the nearby international border.

The U.S. part of the Salmo pack's home range falls completely within national forest lands. Besides extensive road closures to protect grizzly bears and caribou, the Colville National Forest has been a national model for collaborative and progressive timber harvest projects, including efforts to restore many areas of the forest to historic (pre–European-American settlement) conditions. Forestwide standards initiated in 1994 in the Colville and other national forests in the interior Pacific Northwest include prohibiting cutting of any trees over twenty-one inches in diameter to ensure that logging operations also promote the development of the larger trees typical in the past. A bright spot in the world of conservation and rural economics, numerous projects on the Colville have been developed and carried out through a formal collaboration between the USFS and the Northeast Washington Forestry Coalition, which consists of representatives from regional conservation groups and local timber interests as well as other business and forest stakeholders. Many of the collaboration's forest restoration projects and timber harvests have been designed to thin dense stands of trees that have grown up in the wake of fire suppression, creating more open conditions which were the historic norm over much of the area. Other operations have focused on promoting regeneration of aspen stands that were being choked out by conifers and not regenerating in the absence of recent fire. The mosaic of forest stand structures in the area has created a wide variety of habitats for wildlife from migratory birds to resident big game. I was curious to understand how the Salmo pack was using this varied, remote, and rugged landscape.

I walked many miles of closed roads, explored decommissioned logging roads that dead-ended in old cutblocks now overgrown with young trees and brush, and explored stream corridors with eighty-year-old western red-cedars nestled among the massive stumps of their ancestors, shading patches of snow still lingering after an exceptionally big winter. I found signs of the pack all across the landscape: tracks running back and forth along the roads and bone- and hair-filled scats. In the first three days in the field I discovered the remains of several moose and a single deer that the wolves had likely killed and consumed. The deer carcass was just off a gated logging road that the wolves had used. I found one moose carcass in a deep ravine close to the confluence of two fast-moving streams. The second was up on a steep south-facing hillside in a patch of open forest which looked like excellent winter habitat for moose. I watched two black bears over those initial days. The first was a male rubbing on an old post set in the ground at a road junction, a common scent-marking behavior which males engage in most frequently in the spring and early summer leading up to the breeding season. The second was feeding on grass and forbs in a patch of open Douglas fir forest.

Despite the quantity of wolf sign I saw during those first several days, none of it appeared very fresh. A heavy rain fell during the last of those days, and in the morning I set out with renewed energy to determine where the pack might have shifted its activities to. I traversed miles of roads that I had walked previously, hoping for new sign that I could follow but saw none. Finally, with some trepidation given my propensity for epic bushwacking battles in the Selkirks, I left the road to complete a large circuit over a ridge, into an unroaded drainage and hopefully back up the next ridge to a decommissioned old road which I could follow back toward my camp. Hours later, in the bottom of the ravine, I was about to abandon the last portion of the route and attempt to shortcut back to the road when I noticed a well-defined game trail leading up out of the ravine in the general direction I was hoping to go. I was pleasantly surprised to find fresh wolf tracks on it from an animal that must have passed this way during the previous night. The trail quickly collected the tracks of several more animals as it climbed. This was what I was looking for. With more excitement now, I followed the tracks into a selectively cut block of trees and up along the edge of the cut block with the denser forest at its edge. Eventually the tracks cut into the forest. Following as best I could, picking out tracks and estimating how a wolf might travel through the landscape, I began to descend the south-facing slope toward a small stream. In a patch of bare soil multiple fresh tracks going in several directions caught my eye. I paused and carefully inventoried the surrounding area. A little farther down the slope I picked some wolf hair off of a shrub which hung over the game trail and beyond that found several oval depressions in the forest litter also with wolf hair in them—beds.

ABOVE: Several years before this photo was taken, this forest patch was thinned and then burned to help recreate conditions more similar to historic stands in this part of the Selkirks.

OPPOSITE ABOVE: Crowded stands of conifers such as this are typical in many forests in the Selkirks and Columbia Highlands as a result of past logging operations and fire suppression. Such forests have limited value for hoofed mammals and other wildlife species because of a lack of understory plants and limited structural diversity in the even, aged trees.

OPPOSITE BELOW: Numerous roads in the Selkirks are closed to protect endangered grizzly bear and caribou populations. These road closures also greatly increase the security for wolves by reducing human use in these areas.

I continued down to the stream and onto the old road I had been aiming for to complete my circuit. That evening I set up a blind on the end of the road on a grown-over logging landing just below the slope where I had found the hair and beds. A pair of ravens lingered in the forest nearby, and I wondered if they had a nest in the trees there. Wolves and ravens are well known for their strong association with each other; in fact, a pair of ravens had set up shop close to the Diamond pack's den. About an hour after my arrival at the blind location, the ravens, which had been calling from high up in the trees, moved down in the forest canopy and closer to the road, drawing my attention. I followed them by ear as they moved through the forest, coming yet closer to the road. A stocky wolf, classically colored grizzled gray with a cream-colored muzzle, trotted across the road and disappeared into the forest on the other side, apparently heading in the direction of the beds I had discovered earlier.

Over the next several days I monitored the area from the blind. Howls of a lone animal emanated from the forest above several times, responding to or eliciting a response from multiple animals off in the distance. On my last morning there, the howls of several animals echoed out of the forest above the blind. That afternoon, on the six-mile route out from my camp I discovered fresh tracks and scats of wolves along the way, even right by my parked truck at the gate. I had a strange feeling that I hadn't been the only one doing research in the area.

This wolf is a member of northeastern Washington's elusive Salmo pack.

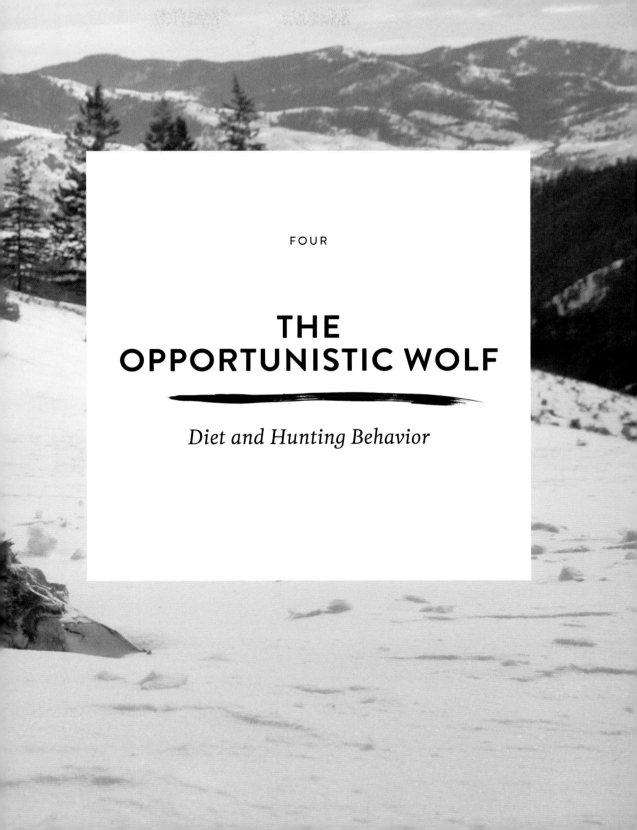

FOUR

THE
OPPORTUNISTIC WOLF

Diet and Hunting Behavior

THE FIRST HINTS OF A BRIGHTENING eastern horizon heralded the end of a cold, clear September night in the Selkirk Mountains on the Washington-Idaho border. A waxing moon cast pale light over the rolling mountainsides and dense forests of pine, Douglas fir, and larch—the territory of the Diamond pack. State wildlife biologist Jay Shepherd and I had spent the day before out in the field in the vicinity of the two wolf packs he was monitoring. This was my first trip to the Selkirks to understand how wolves in this area made their living. We had picked up the signals of two radio-collared wolves in the Diamond pack near a large meadow system; fresh tracks of adults and pups caught our attention at the edge of one of the meadows. That evening I returned to camp in the area and explore and photograph wolf habitat.

My alarm was set to go off about a half hour before dawn, giving me just enough time to pack up my camera equipment, eat a quick breakfast, and be out at first light. But this morning, I didn't need an alarm. I woke to the sound of a chorus of wolves. Low mournful howls, mixed with the shorter yaps of pups, filled the still, dark air, bringing a shiver to my spine.

Though it is notoriously hard to gauge location, distance, and numbers from wolf howls, my sense was that I was hearing most or all of the pack and that the howls were coming from the forest adjacent to the two-mile-long meadow system I was planning on exploring. I quietly packed up and set out, wandering through the forest toward the meadows that flanked the stream running through the valley. On a game trail along the way, I found older tracks of wolves and the fresh tracks of moose. Eventually several fresh sets of wolf tracks appeared on the trail, clearly made after the frost had settled on the landscape the night before.

Carefully I continued down to the meadow. There, losing the tracks of the wolves where the game trail petered out into the dense forest, I moved slowly along the forest edge just inside the trees. The sound of a duck quacking around a bend ahead hinted at open water, possibly a beaver pond. Something had disturbed the duck. Slowing further I eased my way around the densest clumps of trees until I could see the next section of meadow. Indeed a large pond glimmered through the tree branches in the low morning light. Movement caught my eye, and the outline of a large moose came into view as the animal waded farther out into the pond.

PREVIOUS: A gray wolf feeds on the carcass of an elk. Northwestern Montana.

Now in a location with a less restricted view, I was able to watch the adult female moose browse for over an hour. During this time a second female moose, this one with a calf, popped out of the forest on the far side of the clearing, about a quarter mile away. They browsed briefly on the willows and other shrubs before retreating back into the forest, the calf following its mother.

An adult cow moose in the Pacific Northwest weighs about 650 pounds. Watching the first moose feed, her long legs submerged in the water, I thought it odd that she, too, didn't have a calf, as her size indicated she was a mature animal. The thought came and went as I marveled at the huge creature not fifty yards away. Eventually, with the day warming, the moose moved off into the forest. I took this as my cue and ventured forth into the meadow to have a better look.

A large beaver dam built of sticks and mud, spanning the one-hundred-yard width of the clearing, had created the pond. Water seeped through the dam all along its length, making the meadow below it wet and mucky with the dam itself being the only relatively dry ground to cross. Wolf tracks of various ages, going back and forth along the dam, indicated that this was a well-used travel route. Such a spot could be a productive location during a night of hunting, as moose, beavers, and waterfowl are all on the menu for wolves in our region.

I backtracked to where the wolf tracks appeared by the beaver dam and cut into the forest there. Just inside the thick growth along the very edge of the meadow was another game trail, again with the tracks of wolves. I followed it up and away from the pond and meadow, finding several wolf scats filled with moose hair as I went. The trail came to a small clearing in open lodgepole pine forest. There I found several more scats, a variety of bone fragments, clumps of moose hair, and a complete mandible of a juvenile moose, painting a typical picture of predation from several months before. The broken fragments of leg bones were further indication the moose had been fed on by wolves, whose powerful jaws allow them to crack such large bones. The overall size of the jawbone and structure of the teeth indicated the moose had been just a few months old.

It had likely been killed by members of the Diamond pack in the early summer, a time when young moose are still very vulnerable to wolf predation. I recalled my curiosity about why the moose I had been watching didn't have a calf with her and wondered if I might have just found the answer. Though I never saw wolves that day, the ecological story written into the landscape hinted at the influence of these dynamic predators.

The most profound way that wolves interact with their environment is through their stomach and their efforts to fill it. As a social predator specializing in chasing prey rather than ambushing or stalking it, wolves deeply influence the behavior of their primary prey species, large herbivores who in turn play a powerful role in ecosystems. Similarly the quest to obtain these large prey, and to consume them efficiently, has been a primary driver of the form, behavior, and social structure of wolves themselves.

At the same time, and perhaps as a result of their social nature and exceptional global distribution, the diet of wolves and their foraging methods are characterized by opportunism, flexibility, and adaptability. In various parts of the Pacific Northwest and at various seasons, the primary food in their diet ranges from moose to salmon. Similarly their hunting methods vary from testing groups of a hundred or more elk in a large meadow system in the Rockies to traveling the beaches at low tide looking for whatever the ocean might provide, or swimming the open ocean to remote islands to hunt seal pups. They will scavenge carcasses of animals they didn't kill themselves, chase other carnivores off their kills, frequent garbage dumps, ambush beavers where they come out to feed at night, stalk and pounce on meadow voles, and occasionally eat berries when they are in season.

HUNTING METHODS: SOCIAL AND ADAPTABLE

Wolves will hunt deer or larger prey alone or with other pack members. In a pack setting, often only the breeding pair will initiate or lead a hunt, and even a single wolf is capable of killing an elk or moose in the right conditions. As is typical of social carnivores around the world, hunting group size increases as the average body size of prey increases and as the hunting environment becomes more open. Conversely, hunting smaller prey in forested environments tends to require fewer wolves.

The first step for wolves in a hunt is to locate potential prey. This often involves wide-ranging travel to find prey or the scent of prey to follow. When they identify areas heavily used by prey, they will focus their search efforts there. At other times they encounter prey opportunistically in the landscape. Having more animals on the search likely increases a pack's chances of success. If the prey is unaware of the wolves' presence, they will attempt to close the distance undetected. Once a potential prey animal recognizes the wolf, the animal will either stand its ground, flee, approach, or charge the wolf or wolves. Wolves want prey to flee and usually stand down to a charging animal. An animal that stands its ground is often left unmolested, or a wolf may charge in an attempt to get it to flee. With smaller or weak prey (fawns or smaller), this initial rush is a fast attempt to immediately subdue the animal. If a large animal begins to flee, a chase ensues. During a chase wolves are testing the vigor of the prey or may attempt to move the prey into terrain where maneuvering is more difficult for it. When chasing a group of prey, wolves often seek out weaker members of the herd that may tire more easily, or animals that become separated from the group. Chases are generally relatively short, typically less than a half mile and often no more than a couple hundred yards. On rare occasions, however, chases may extend to three to five miles. All the chase sequences

I have observed through tracks have ended after just a few hundred yards or less of running. Yet even under a wolf's most ideal hunting conditions, the vast majority of such chases end unsuccessfully for the predator.

To kill big game, wolves employ several strategies. They will attempt to grab the neck or hind quarters of fleeing animals. Elk cows and calves are more likely to be attacked at the neck, whereas bull elk are more likely to be attacked at the hind quarters. Multiple bites to the hind quarters disable the animal, allowing the wolves to bring it to the ground. Killing bites, usually to the neck, may crush the windpipe, killing the animal from asphyxiation. However, sometimes wolves may begin to consume an animal before it has died.

Caching of extra food is common for wolves in many areas. They may carry off and bury part of a carcass or regurgitate ingested food into a hole dug for this purpose. Wolves dig such holes with their front feet and often use their nose primarily to bury the item

This shallow excavation is the remains of a wolf's cache. Note the remains of a young deer pelvis adjacent to the dig. Selkirk Mountains, Washington.

after depositing it. After a single feeding, wolves can regurgitate multiple times into multiple caches or directly to their young. For caching they generally carry the food some distance from the main carcass, likely to conceal it from scavengers or even from their own pack mates. In comparison, solitary carnivores such as mountain lions or bears will attempt to conceal an entire carcass.

Most hunts involve simple chases with younger animals following and participating as much as possible or not at all. Hunts often do not involve all pack members. Research suggests that a pair of wolves is more effective at killing hoofed mammals than an individual wolf, but the addition of more wolves beyond two, though helpful in some circumstances, does not consistently increase their effectiveness. One reason efficiency may not go up with more than two animals might be that there is only so much room around a fleeing animal for multiple wolves to attack. Another factor might be that as young wolves mature and become more efficient hunters, they often disperse, taking their prowess with them. This leaves the breeding pair the burden of feeding remaining pups and subadults. In contrast to wolves, the hunting efficiency of African lions has been documented to go up with increasing numbers. In African lion prides, adult females live together for many years, which gives them the opportunity to develop more coordination in hunting endeavors.

Social hunting may also help in the training of younger wolves. Acquired social tradition and methods successful in specific locations could be passed on, though demonstrating this empirically is difficult. Animals raised in captivity have been documented to effectively kill prey as large as elk, and translocated animals quickly learn to kill novel prey in new environments, implying that traditional training is likely not a requisite part of wolf hunting methods.

Wolves have a dynamic population structure, members and locations of packs change often, and young animals regularly disperse widely to new environments. Adaptability to novel hunting environments and an ability to kill successfully despite lack of social coordination are important attributes that have likely contributed to their ability to reoccupy landscapes where they had been removed. While more experienced, local animals and groups of wolves may have an easier time acquiring prey in a given location than a recent arrival, recent arrivals should be able to figure it out fairly efficiently.

On a sunny winter day Ray Robertson and I were out trailing a member of Washington's Lookout pack. We followed the large tracks over a ridgeline and enjoyed some pleasant skiing down the north side of the ridge, following tracks to an area above a small bench. From here we could see the wolf's tracks approach the remains of a deer. The carcass was partially drifted over with snow, just the branched antlers and part of the rib cage sticking out. We came to the carcass carefully. From the tracks we could see the wolf had approached, lingered briefly, and then moved off at almost a right angle, indicating that it had likely traveled directly to the carcass to inspect it. There was no sign that it had fed on anything, and indeed the carcass was picked clean down to the bones.

It is not uncommon to find evidence of wolves revisiting a carcass several days after a kill, especially if their initial feeding was not extensive and edible bones and hide remain, or if they have not made subsequent kills. However, a pack of wolves may spend as little as three hours at a kill site of deer-sized prey after which nothing but the stomach contents would remain. One study showed an average of less than twelve hours at a site for a deer-sized animal. Larger prey may take longer to consume, and of course a lone wolf or smaller pack will take longer to consume a carcass.

Carcasses of large prey are consumed in a fairly predictable fashion. The body cavity is opened and the internal organs are generally consumed first, followed by the large muscles of the hind legs, then the smaller muscle groups, and last the bones and hide. If possible the brain case will be opened. The first feeding on a carcass lasts about thirty minutes to an hour, and a single wolf may consume up to twenty-five pounds at the first feeding. Within a pack, the social hierarchy determines feeding order: in larger packs, less dominant animals will wait until the first wolves feed and go off to rest before feeding themselves. During the pup-rearing season, after an initial feeding, a wolf may return promptly to the den or rendezvous site to feed pups and a breeding female. At other times a wolf may rest, often close to the carcass, for several hours. After this, a second feeding may occur followed by caching of any remains. Often only the bones remain after a second feeding.

Inspection of a wolf-killed ungulate carcass tells us a lot, not only about the behavior and ecology of wolves and their prey but also about the wide array of other animals—from Stellar's jays to bears—that make use of the bonanza of food. A wolf pack's speedy consumption of a carcass, in part driven by competition among packmates, is likely also an attempt to minimize loss to other competing species. What wolves do not consume immediately will often be vigorously defended from bears, coyotes, and smaller scavengers. One major culprit is ravens. Up to 133 ravens have been observed on a wolf-killed moose carcass, and they can remove up to eighty-one pounds of meat a day according to estimates. A pair of wolves might lose half of a moose carcass to scavengers, while more than two wolves, because of their faster rate of consumption, lose a significantly smaller portion of their prey. The larger the prey, the greater the value of the wolves' social nature. Theoretical models of the costs (competition with pack mates) and benefits (increased rates of prey acquisition and efficiency in carcass consumption) of group living for wolves have found that the only energy-saving benefit to group living was in protection from the high level of loss to scavenging ravens! This finding suggests that minimizing the impacts of scavengers stealing prey killed by wolves has likely been an important evolutionary driver for their social nature. Despite being as efficient as possible, wolves still end up subsidizing the diets of a variety of other scavenger species, another important ecological role wolves play in our region.

In its review of the potential impacts of wolf recovery in Washington, the WDFW estimated that a wolf in the state would take an average of 8.5 to 12.6 elk and 14.1 to 20.9 deer or a total of 22.6 to 33.5 wild ungulates per year, the rate varying seasonally and regionally. Research from Yellowstone suggests that winter kill rates of ungulates are likely higher than in summer, both because of increased prey vulnerability in the winter and increased diversity in wolves' summer diet. In coastal environments, where wolf diets include a significant amount of marine mammals, marine invertebrates, and salmon, their dependence on hoofed mammals is similarly reduced.

As with the extra food that wolves provide to scavengers, their ecological impact on ungulate populations does not end with consumption. In fact, their impacts on other carnivores and on the behavior of prey animals can have further, far-reaching effects on landscapes.

ABOVE: *Ray Robertson and I discovered the carcass of a buck mule deer while trailing a member of Washington's Lookout pack. North Cascades, Washington.*

OPPOSITE: *This elk had just been killed by a wolf or wolves within a half hour of when I took this photo. As is typical, feeding had begun by entry into the body cavity below the rib cage. Forty-eight hours after the kill was made, only the stomach contents of the elk and a few tufts of hair remained at the location. Northwestern Montana.*

PREY SELECTION

While exploring a large meadow system in the Salmon River Mountains of central Idaho, Emily Gibson and I investigated the remains of two adult bull elk within about three hundred yards of each other. They were located alongside separate forks of a shallow oxbowing stream just above where the two forks joined. Emily found the first while going to fetch water at dusk, noting the bones and rank smell among the waist-high willow thickets. Such carcasses can be dangerous primarily because bears may be aggressive in defending the food if they take your actions as attempts to usurp their treasure. (Wolves typically retreat from such carcasses at the advance of a human.) Emily quickly retreated to the spartan base camp we had set up for tracking and photographing the pack.

The next day we returned to a scene that was typical of such sites in this part of the Northwest. The carcass lay adjacent to a stream bank in the brush (sometimes carcasses are actually in the stream), suggesting that the wolves might have captured the elk while it was feeding on the willows or chased it into this terrain, which slowed the animal down and allowed the wolves to overtake it. Much of the flesh had been consumed, but the skull and the leg bones were intact. Flies swarmed in grass patches soaked with blood, and the smell was overpowering in the growing August heat. Wolf tracks and scats were plentiful around the area. We found scattered collections of bone fragments, another signature of wolves feeding on a large carcass, putting their powerful carnassials and molars to work to break and consume even the bones of their prey. The bull elk's velvet-covered, growing antlers, a blood-rich and nutritious meal, had been chomped down considerably. Emily discovered the second carcass later that day while attempting to cross the other stream, which was flooded by a network of recent beaver dams. This carcass appeared to have been on the ground for about the same amount of time as the first. Together these two carcasses explained the unusually high amount of wolf activity in the immediate area, a place we were unaccustomed to being such a hub.

Why did wolves select and kill these two bull elk, as opposed to any others? Both appeared to us to be healthy adults, about six or seven years of age with no signs of disease or handicap. During the summer months bulls will often group together into what are referred to as bachelor herds. We surmised that the pack had encountered such a herd and had taken advantage of the terrain to kill two of them, possibly at the same time, or very nearly so. The wolves had then moved their pups to a secure location near the sudden bonanza of food.

How and why do wolves select particular individual prey or one species of prey over another? The answers are multifaceted. Across most of the Pacific Northwest ungulates comprise the vast majority of a wolf's diet. However, the particular species or combination of species varies widely, depending on environmental conditions, body size of prey animals, relative abundance of prey species, relative safety of the pursuit, and relative energy expenditure required to find and subdue prey.

Varied factors affect the vulnerability of prey. Wolves often select a prey species at a higher rate than would be predicted by their relative abundance. During the rut, when male ungulates are exceptionally preoccupied with breeding activities, they tend to be killed more often than females. Females are more vulnerable in late winter when they are close to full term in pregnancy. Very young and old animals are generally more vulnerable. Malnourished animals are more at risk from wolves as their energy reserves are depleted or they take greater risks in their attempts to find food. Any injuries, abnormalities, diseases, or parasites that impede a prey animal's ability to detect, deter, or flee from wolves puts it at greater risk. Conversely, more aggressive animals are taken less often than those that flee. In some studies the age of a young animal's parents have been implicated in the likelihood the juvenile will be killed, possibly because of poorer parenting skills of younger parents.

To reduce the risk of being killed by wolves, ungulates employ a number of biological and behavioral strategies. Prey species have been documented to increase their use of the buffer zones between wolf packs. To protect their young, prey species often seek out habitat for birthing that isn't heavily used by wolves. Hoofed mammals have highly synchronized birthing, a behavior referred to as swamping, which decreases the window of highest vulnerability for juveniles while ensuring that even with high levels of predation some young will survive. Interestingly, prey species' behavior in response to wolves can vary drastically depending on the specific ecological conditions. Some studies have shown that elk select open areas away from forests and obstacles in order to detect approaching wolves, while other studies suggest that elk appear to select forest cover in the presence of wolves, perhaps to decrease the likelihood of detection by the predator.

In a classic hunt by a wolf or other cursorial predator (one adapted for running and chasing), slow, weak, old, or young animals are sorted out of a herd because they fall behind their herd mates and are therefore more vulnerable. In this setting wolves test prey and choose the most vulnerable to pursue, or cease pursuit if they perceive all to be overly fit. Yet wolves are not necessarily limited to killing only relatively weak animals, as we saw with the two bull elk carcasses in central Idaho. Wolves use other hunting methods as well, such as ambushing prey like deer in dense cover. Here the elements of surprise and proximity at the time of detection increase the wolf's chances of successfully subduing a healthy animal.

Wolves adapt to the conditions they find themselves in, opportunistically taking the prey they can access with the least risk to themselves and least energy expenditure. For instance, if elk are easier to find they might receive disproportionate attention even if they are a relatively small population compared to deer. Occasionally a pack seems to have specific preferences, but this may just reflect what is most easily accessible at the least risk. Additionally, experience hunting a certain type of prey might make hunting it more efficient and less risky. Sometimes environmental conditions make otherwise healthy and relatively less vulnerable

animals more susceptible to predation. Particularly hard winters can leave fawns and calves malnourished and more vulnerable to predation even after the harsh conditions of winter pass. Conversely, easy winters leave prey in better condition and less vulnerable to predation.

Moose, which defend themselves vigorously, are generally more dangerous to wolves, and can travel through winter snowpack more easily than deer. In places where moose are the only ungulate species whose winter range overlaps with a wolf pack's territory, it makes sense that wolves would risk taking on these large and dangerous prey. Yet researchers in northwest Montana found moose were selected disproportionately compared to their abundance in winter conditions in areas with much higher deer abundance. Besides selecting prey for overall vulnerability, wolves may instinctively base selection on the likely overall profitability, or net gain, of the hunt. Moose, being much larger than deer, provide more food for wolves per hunt. Another factor that contributes to profitability is the search distance required to find prey. In the same northwest Montana study, researchers found wolves spent most of their time hunting in areas that had the densest deer populations, which would increase the likelihood of an encounter with deer. Moose and elk in the study site were more evenly spread out across the landscape, making the energy required to find them much higher than for deer. However, within the deer wintering area, wolves apparently pursued moose or elk whenever they could, possibly because they represented a larger food source than deer.

Many hoofed animals, such as these elk, employ a herding strategy in open terrain, which reduces the risk of predation for individuals. Wallowa Mountains, Oregon.

Because of their preference for more vulnerable prey, one might expect that wolves would kill livestock whenever they are available, as they are generally less fit, less aware, and less dangerous than wild prey. Interestingly, this is far from the case. While wolves in the Pacific Northwest and in adjacent areas definitely do kill livestock, there are also a great many instances of wolves pursuing wild prey in landscapes that also included domestic stock. Volker and Iris Steigemann, on Cortes Island in British Columbia's Desolation Sound, recounted their experiences interacting with wolves on the small farm they have owned for twenty-five years, including one season when the resident pack had their summer rendezvous site on the beach adjacent to their farm. They told me about discovering several deer carcasses on their property where they raised sheep and chickens. They noted that a cougar once killed a sick sheep of theirs, but in their twenty-five years here, wolves had never touched their sheep, though they had killed chickens from time to time. It is unclear, from a biological perspective, why wolves would not consistently utilize such food sources given their strong drive to take the most vulnerable wild prey available.

In forested environments, deer and other hoofed mammals will disperse across the landscape, to blend in and avoid detection by wolves and other predators. Puget Sound, Washington.

The Pacific Northwest's current resurgence of wolves parallels the resurgence of wolves and other large carnivores across the globe. In all these places wolves are coming into contact with prey that have had no experience with these predators for 50 to 150 years or more. Moose, deer, and elk in locations with recently recolonized populations of wolves have all demonstrated a very steep learning curve in dealing with the threat of wolf predation. Researchers have documented changes in the feeding behavior and vigilance of individual animals as well as shifts in habitat use by prey species after exposure to wolves as a predation threat. For instance, female moose who have had a calf killed by wolves show significantly increased vigilance and abandonment of their feeding site when exposed to wolf howling, wolf scent, or even raven calls compared to moose in areas where wolves are absent. Studies from Yellowstone National Park noted a change in the locations where wolves killed elk since the initial years of reintroduction, with parts of the landscape clearly becoming important hunting grounds and others clear refuges from predation. These shifts in hoofed mammal behavior can have significant ecological impacts. They also further suggest that it is unlikely that returning wolves will completely annihilate unsuspecting prey species under natural conditions.

Wolves (and their evolutionary ancestors) and hoofed mammals have coevolved for millions of years. In her book *The Wolf's Tooth*, on the dynamics of apex predators such as wolves, Cristina Eisenberg quotes poet Robinson Jeffers: "What but the wolf's tooth whittled so fine the fleet limbs of the antelope?" The drive to kill and avoid being killed has shaped the evolution of form, speed, keen senses, and predatory and defensive behaviors of wolves and hoofed mammals in our region today. Through this process, something akin to a biological arms race (as well as a literal arms—or limbs—race), these species have created a fairly equal playing field, a fine balance in capacity between predator and prey, and produced a fascinating array of hunting and defensive strategies for each. In most natural settings, during times of abundance for prey, such as a series of mild winters, their numbers will grow despite wolf predation. In difficult conditions for prey, such as during or following particularly harsh winters, wolves can contribute to significant population decline. This delicate relationship has allowed both predator and prey to survive and evolve over millennia. If a predator's capacity is far beyond the ability of a prey species to manage, that species would disappear over time.

Despite this relationship, some people fear that returning wolves will annihilate deer and elk in the Pacific Northwest. There is precedent for a predator sending a prey species toward extinction, typically involving a novel predator coming into contact with prey species. Numerous birds on South Pacific islands have been devastated by introduced predators such as house cats and rats. Invasive species have been implicated in the extinction of numerous species of birds, fish, and mammals around the globe. A highly modified landscape, which reduces the effectiveness of the defensive strategies of prey species, can also significantly increase their vulnerability to predation. As we see with the enormous human-caused shifts

This black-tailed deer is demonstrating a high degree of vigilance with its head up and eyes and ears directed toward the thick forest, a potential hiding spot for a predator. Olympic Mountains, Washington.

in the Selkirks, wolves may be a significant threat to small groups of endangered caribou that have literally had the forest pulled out from under them. Humans have demonstrated their unparalleled predatory ability in their destruction of wolf populations themselves. The only reason wolves have begun to return to landscapes that we vanquished them from is a change in our behavior as their predator. Wolves evolved to survive in the face of competition and aggression from lions, tigers, and bears, but the organized and highly sophisticated methods used by the human species to kill wolves overwhelmed both their individual abilities to evade humans and the species' prodigious reproductive capacity. When it comes to deer and elk in the Pacific Northwest, however, neither are wolves an evolutionarily novel predator, nor have habitat conditions changed in a way that radically increases their vulnerability. To the contrary, the presence of humans, and our built environment, sometimes acts as a shield for prey species against predation by wolves.

HABITAT, TERRAIN, AND HUNTING SUCCESS

Despite valiant attempts by wolves to take advantage of the most vulnerable prey available, the ratio of successful to unsuccessful hunting endeavors is relatively low, usually less than 25 percent and often as low as 1 to 4 percent. Hunting success is likely highest in winter when prey are most vulnerable, especially in places with significant winter snowpack, where snow conditions are the strongest driving factor in a successful hunt, as in most mountainous terrain in the Northwest.

Deep loose snow impedes the travel of deer more than wolves, leaving them more vulnerable to predation. In many of the western and low-elevation parts of our region, snow conditions are not a strong driving factor in typical winters, or not in quite the same way. On Vancouver Island where large portions of land have experienced recent clearcutting, deer seek out remaining patches of mature forest during winter, especially when there is snow on the ground, for both shelter and for access to browse. This congregation of deer into a small part of the landscape makes them more vulnerable to wolves who have an easier time finding them.

This Rocky Mountain elk was killed in the uneven ground adjacent to an oxbowing stream in the middle of a large prairie system in the Salmon River Mountains of central Idaho. The uneven ground and vegetation obstacles may increase the ease of capture.

Besides snow, a variety of other terrain features come into play. For a number of years I co-led wolf-tracking expeditions in central Idaho for Wilderness Awareness School. Since the school started these expeditions in the late 1990s, many of the wolf-killed elk carcasses we've discovered have been in riparian vegetation. We also find elk carcasses in the large open prairies that stretch across the mountain valleys, as well as in the forest and forest's edge, but much less consistently here

than in riparian areas. Douglas Smith, director of the Yellowstone Grey Wolf Restoration Project, reported that their project found clusters of wolf-killed elk in riparian areas and ravines. In an interview with Cristina Eisenberg, he described watching a wolf chase an elk into and through such a ravine, noting that the wolf effectively closed the distance to the elk in that more difficult travel terrain and that numerous carcasses were documented in that particular ravine.

Sometimes such clusters of carcasses have been attributed to wolves ambushing prey that are unable to see them coming in a ravine or around obstacles. Smith said, however, that wolf-elk encounters always involve a chase and that they rarely end where they began, implying that the carcass location says more about the type of terrain that lends itself to a successful kill than about where the initial encounter occurred. Further, while elk and other ungulates do use sight to detect predators, their sense of smell is by far more important to them. Obstacles to their view would not necessarily keep them from detecting oncoming wolves by scent.

Wolves will increase their effectiveness by spending more time hunting in areas where prey are more vulnerable or possibly by attempting to drive prey into such locations. What such terrain traps look like varies, depending on the prey species. While elk are generally believed to have evolved primarily as an open country species, deer are clearly evolved as a forest species that effectively uses brush and ground obstacles to conceal themselves. However, a study of wolf hunting behavior in northwest Montana also discovered that they were more successful in killing deer in areas with larger amounts of ground cover, in part possibly because they could approach prey more closely before detection than they could do in the open. Wolves also made more kills in ravines and on lake ice than would be predicted by the amount of such terrain in the landscape. In the case of lake ice, it is very likely that these prey were chased onto such terrain rather than being discovered there, an example of wolves taking advantage of the terrain to increase their effectiveness.

Research into the diets of wolves on the central coast of British Columbia found that wolves inhabiting coastal islands had significantly more marine food sources and salmon in their diet than did mainland coastal wolves. Researchers attributed this diet variation to decreased abundance of deer, increased access to marine environments, and in the case of increased salmon consumption, decreased competition with grizzly bears who were less abundant or absent on islands.

While more efficient in some landscapes than others, in any environment where prey are available, wolves have demonstrated an ability to procure them. The expansion of wolves into landscapes where they have been absent for as much as a century is a testament to their ability to adapt to and successfully hunt in novel environments, including human-created landscapes such as the large swaths of industrial-scale clearcuts which Washington's Diamond pack makes use of for hunting.

MORE THAN MEAT ON THE HOOF:
THE OPPORTUNISTIC WOLF

What wolves eat in a particular area is probably the most significant way that this carnivore influences the ecology of its environment. Their diet is also often the largest source of conflict with human interests. In places where wolves were eradicated decades ago, there is close to no historical data on their diet. Even if there were, this information would be of little value as ecological shifts in the region would likely alter the foraging behavior and diets of wolves today.

As with most predictions in complex systems, attempting to extrapolate what the diet of wolves will be in one location from what they eat somewhere else can be challenging. On Vancouver Island, in locations with both deer and elk, deer comprised the majority of the wolves' diet. In much of the Rockies, where elk are abundant they become the primary prey for wolves. In Banff National Park, elk are the most abundant ungulate and most common prey for wolves. In Yellowstone, elk comprise 88 percent of diet despite availability of eight species of ungulates. However, in one study during the winter in northwest Montana, wolves primarily preyed on deer in areas that also had resident moose and elk. Here deer comprised 83 percent of their diet compared to 2 percent moose and 14 percent elk. Further, wolves here focused their hunting efforts on the more limited winter range of deer, which was more efficient than primarily pursuing elk and moose who ranged more widely in the winter. In winter habitat containing more clustered elk populations, wolves would likely have a diet that reflected this.

While hoofed mammals are generally the primary food source for wolves, they are by no means the only thing on the menu. Wandering around the mountains above the Methow Valley in Washington, wildlife trackers Gabe Spence, Brian McConnell, and I came across a large scat on a ridgeline game trail. This location and the scat's large size and smooth tubular appearance told us it was the scat of a wolf. Using a stick Gabe carefully broke open the scat, curious about what our canine neighbor had been eating lately. At the time, Gabe was working on a beaver relocation project, live trapping and moving nuisance beavers to areas that had historically had beavers before the era of intensive fur trapping a century and a half ago. Gabe inspected the dense fluffy hair that comprised a large part of the scat and pronounced it the under-fur of a beaver, something he had become very familiar with.

In numerous diet studies across the continent, wherever the two species overlap, the North American beaver (*Castor canadensis*) has often been the most important non-ungulate food source for wolves, a fact also documented in parts of our region. Their relatively large size makes beavers an attractive meal for wolves and other large predators. Wolves travel watercourses looking for beavers or wait in ambush at dams where beavers predictably cross overland and are most vulnerable.

Besides actively hunting prey, wolves are effective opportunistic scavengers. The pack of wolves that discovered this elephant seal carcass on the beach returned daily to feast. Clayoquot Sound, British Columbia.

Northwest beaver populations plummeted in the 1800s because of the fur trade. The first laws protecting them regionally were enacted in the 1890s. Starting in the 1930s, beavers were translocated in an attempt to reestablish the species across the region. The resurrection of beaver populations here has had profound ecological benefits for a multitude of species such as song birds which nest in the increased riparian vegetation created by water impounded by beaver dams. Beavers provide a wide variety of ecosystem services including helping moderate stream flows and raising water tables.

In Oregon and Washington west of the Cascades, beavers are once again abundant. As they are for wolves, beavers are an important secondary food source for mountain lions in many areas. A study in western Washington showed beavers were

ABOVE: A raccoon forages for sandfleas on the beach. In diet studies from coastal British Columbia, raccoons and river otters have been found in unusually high numbers in the diets of wolves compared to elsewhere in North America.

OPPOSITE ABOVE: Beavers, once close to extinction across much of our region, are again abundant in much of the Pacific Northwest. Often an important secondary food source for wolves, they are hunted when traveling overland between water sources such this one crossing one of its dams in western Washington.

OPPOSITE BELOW: A river otter scent marks on seaweed in the inter-tidal zone. In coastal locations, river otters often end up on the menu for wolves. Vancouver Island, British Columbia.

mountain lions' second most important food source, after deer, but more important than elk. At 22 percent of the mountain lion diet, beavers comprised an unusually high percentage here compared to most other environments the two species cohabitate, which might also become the case with wolves in this part of the region. One diet study of wolves on Vancouver Island noted that beavers were consumed in the winter and spring, an atypical time compared to areas with deep snowpack, where beavers are less accessible to wolves during the winter months and a significant prey item only in the summer.

One study of the wolves along the British Columbia coast found a fairly low proportion of beavers possibly because of a sampling bias toward marine shoreline environments and away from forested interiors, or possibly because of low beaver population in an area that is poor habitat for the species. However, in this coastal landscape, researchers found an unusual abundance of another species—salmon. As we'll explore next, salmon have profound impacts on the behavior, habitat selection, and ecological role of wolves in the coastal portion of our region.

Beyond hoofed mammals, beaver, salmon, and other marine food sources, the items that make their way through the digestive system of wolves are quite diverse, from huckleberries to wild turkeys, voles to grizzly bears, and barnacles to whales. Though every diet study of the species in North America has found that wolves consume animal tissue almost exclusively, several studies from our region or adjacent areas have documented very minor consumption of plant matter by wolves, primarily wild berries.

The most unusual regional food account that I have come across is from Volker and Iris Steigemann of Cortes Island, British Columbia. They documented wolves eating apples from under the trees in their orchard and squash out of their garden, a behavior much more typical of the smaller, omnivorous coyote, which are absent from the islands in Desolation Sound, as they are from Vancouver Island and much of the mainland coast of British Columbia. While berries and other plant matter is undoubtedly inconsequential in the diets of wolves in our region, their consumption is a reminder of wolves' omnivorous ancestry and potential future evolutionary adaptability.

This humpback whale carcass, which washed up on an island beach off the west coast of Vancouver Island, showed signs of being fed on by wolves.

Wolves are not proud when it comes to where they find their food. While I was researching wolves on the central coast of British Columbia, there were reports of wolves howling at the dump outside the small coastal town of Bella Bella. One pack in Washington State apparently made a habit of visiting the scrap pile behind a butcher and taxidermy shop, making off with the distinctively sawed bones, which were later discovered by researchers at the den site miles away.

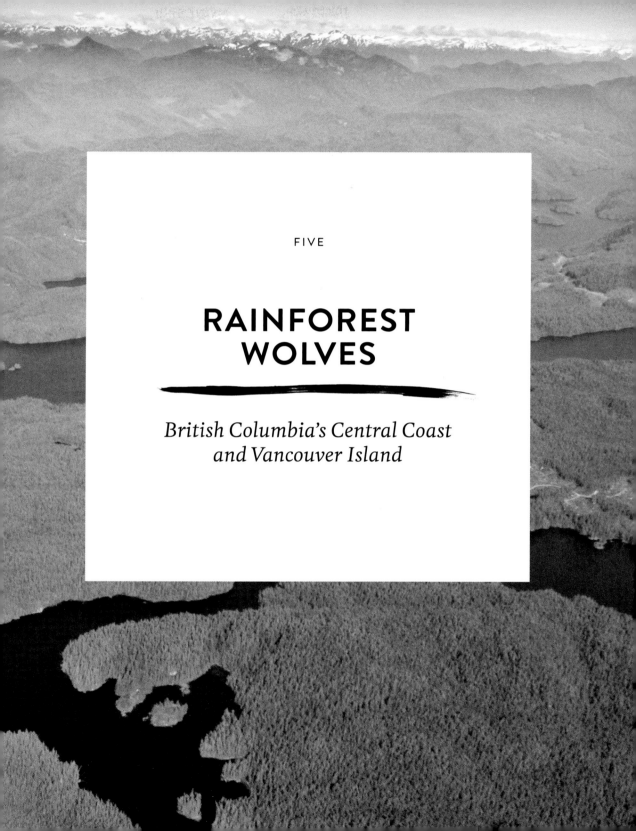

FIVE

RAINFOREST WOLVES

*British Columbia's Central Coast
and Vancouver Island*

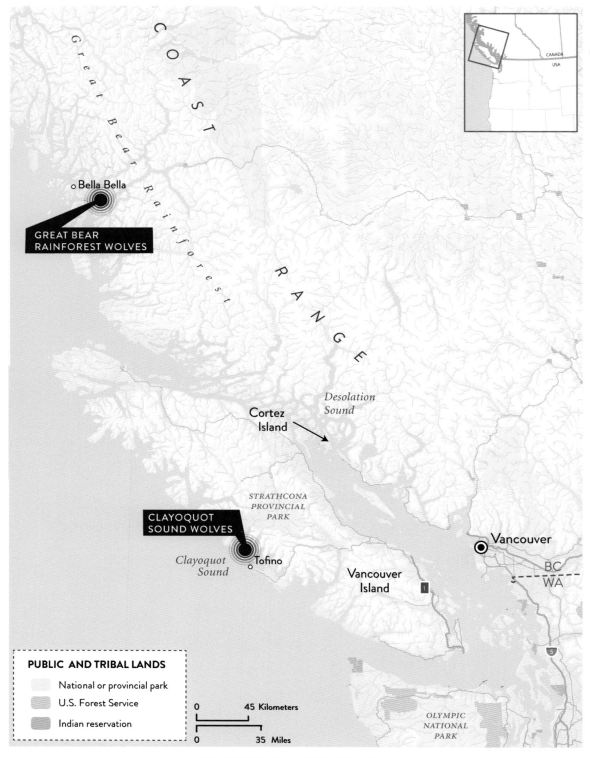

COAST

Great Bear Rainforest

RANGE

o Bella Bella

GREAT BEAR RAINFOREST WOLVES

Desolation Sound

Cortez Island

STRATHCONA PROVINCIAL PARK

CLAYOQUOT SOUND WOLVES

Clayoquot Sound   o Tofino

Vancouver Island

Vancouver

BC
WA

OLYMPIC NATIONAL PARK

CANADA
USA

PUBLIC AND TRIBAL LANDS

National or provincial park

U.S. Forest Service

Indian reservation

0   45 **Kilometers**

0   35 **Miles**

BRITISH COLUMBIA COAST

FIRST THE CROWS CAME, flying around the bend of the coastline a quarter mile to the north, their black bodies in stark contrast to the light sand and hazy breakers as they alternately flew, landed, and hopped along the beach. Then came the wolves, two of them, trotting down the beach toward the small stream that meandered out of the dark ranks of Sitka spruce, western hemlock, and western red-cedar trees. The tide was low, and the stream spread out as it made its way across the wide beach toward the frothing waves of the Pacific Ocean. I was hunkered down in the foredune where the stream and forest met, waiting. To the northeast sunlight had just begun to filter over the tall peaks that run along the spine of Vancouver Island, cutting through the morning fog. A steady onshore breeze assured me that my scent would not make it to the foraging wolves. Two days of scouting, a wakeup at three in the morning, and an hour-long trek in the dark along the beach had paid off. These were the first rainforest wolves I had ever seen.

The previous morning I had picked up the tracks of a pair of wolves trotting down the beach and followed them to this stream and beyond before the tracks circled back and disappeared into the forest along the banks of the creek. Upstream, above where the high tide scoured the soil twice a day, I had found layers of wolf tracks, clearly indicating a well-used travel route. Farther on, the wolves cut onto a network of well-worn bear trails. Moving slowly and more attentively, I picked out the wolves' tracks along a trail created primarily by black bears, littered with broken salmonberry stalks, munched skunk cabbage, and occasional piles of berry-seed-filled bear scat. Eventually the wolf tracks petered out, and the maze of potential travel routes made anticipating their likely path impossible. I sat on a large fallen spruce log taking in the scents of the forest, the sun dappling through moss- and lichen-clad branches, and the distant sound of the surf.

Eventually I returned to the beach and retraced the wolves' trail for much of the rest of the day, inspecting their foraging activities along the beach and scattered dunes. The remains of a raccoon and several wolf scats containing some sort of shelled marine creature made me curious about what exactly was on the typical menu for this beach-combing pair of wolves. With low tide around first light, I had guessed that the wolves might be back the next morning to forage along the beach again.

PREVIOUS: A wolf peers out of the high grass in a wet meadow. Central coast, British Columbia.

As the wolves continued their approach, I could make out the reddish high-lights in their pelage that I had heard were typical of many coastal wolves in British Columbia. They reached the stream close to where it joined the ocean, about two hundred yards from my location. The larger of the two wolves began to walk up the stream, sniffing intently, while the other lay down in the sand and scanned the beach. Half a dozen crows walked around near the wolves. Soon the foraging wolf found something of interest in the sand and began to dig earnestly. The second wolf looked on with moderate interest but did not rise. After thirty seconds of digging, the paws came out and the muzzle went in. I strained to make out the details, but I was too far away to see what the wolf was excavating, either through binoculars or my telephoto lens. The smaller wolf had not risen and did not appear to expect a piece of whatever its companion was digging for, most likely something small like a clam. A little bit of chewing and whatever had been buried in the sand was gone. The smaller wolf stood up and walked over to the excavation where it paused to sniff briefly.

The larger wolf approached, and they sniffed each other. The smaller wolf turned and began trotting downstream toward the forest (and me). The larger one also moved in my direction but walking and wandering, its nose close to the ground. When the first wolf had closed half the distance to me, the second broke into a trot as well, following the first. The crows, which had scattered, came together on the drift logs across the stream, where I anticipated the wolves would likely go. Both wolves were together now, fifty yards away. They disappeared behind a small sand dune at a bend in the stream. I turned my attention to where they should pop out once they had crossed the water.

Nothing. The crows moved closer to me, one perching within twenty yards. Five minutes, ten minutes, nothing. What happened to them? After disappearing they should have popped back into view in seconds. I sat and waited for a half hour. The crows were long gone. I carefully stalked out to a vantage point so that I could see the spot where the wolves had disappeared. I found their tracks and followed them along my side of the stream—the wolves had not crossed it. Both thrilled and disappointed, I pieced together what had happened. Their gait had stayed at a steady trot when they passed closest to me. The fact that they had not stopped (and indeed their tracks showed no sign that they had paid any notice to my location) told me that they had been oblivious to my presence—just as I had been to theirs. Right along the stream was a stretch of sand about four feet wide that I hadn't been able to see as it was tucked against the four-foot-high stream bank. As I was watching the nearest crow, the crow was watching the wolves pass by me. They had come within fifteen yards, each of us unaware of the other.

I packed up my gear and set out along their tracks, first following them upstream along the creek but then back out into the sand and dunes south of the stream. The morning sun had been burning off the fog and haze, but now after thirty minutes on the trail I could barely see as a thick blanket of fog pushed in from the ocean. I abandoned their trail, realizing that all I could expect in these conditions was to spook them. When I returned to the area where the wolf had been digging, the ocean had washed away any evidence of what it might have eaten.

The Pacific Northwest is home to the largest remaining temperate rainforests in the world. These massive conifer-dominated forests, which cloak large parts of the western portion of the region, comprise a biologically rich ecosystem that spans the international border and includes four UNESCO (United Nations Educational, Scientific and Cultural Organization) biosphere reserves and four national parks.

There is more to a rainforest than just the trees and rain. In addition to inviting large amounts of precipitation and a moderate climate, the proximity of the rainforest to the Pacific Ocean brings many other gifts. Northward starting in Puget Sound, the coastline is a convoluted maze of large sounds, archipelagos containing hundreds of islands, and glacier-carved inlets. British Columbia's coast stretches approximately 486 miles as the crow flies between Washington and Alaska, but the actual coastline is 14,579 miles as the ocean wraps around islands and inlets that bring salt water many miles into the continent. Beyond the reach of the salt water, thousands of streams and rivers host runs of seven different salmon species that carry marine nutrients up to nine hundred miles inland.

The vast and inaccessible lands of British Columbia's central coast include the largest tracts of intact rainforest in the region. Although threats to the area's ecological integrity continue to loom over this landscape, regional and global efforts at protection made great headway starting in the 1990s. At that time, conservation groups coined the name Great Bear Rainforest as part of their campaign to draw attention to the value and plight of the region. The name has stuck, and the area has become a rallying point for conservation initiatives as well as a growing eco-tourism destination. The rugged Coast Range guards against easy approach from the interior, while the confusion of inlets and islands makes for a cumbersome approach by water. Most of the Great Bear is completely inaccessible by road. Forests of spruce, hemlock, and western red-cedar crowd down to the salt water's edge and up to the glaciers along the higher peaks of the Coast Range.

The rainforests of the central coast are home to a number of wildlife species that have become rare or that have disappeared altogether from rainforests in the rest of the Pacific Northwest. Marbled murrelets nest in the ancient trees and fish in the salt water. Spotted owls hunt the dark forests. Grizzly bears fatten on salal berries all summer and gorge themselves on salmon runs that fill freshwater streams up and down the coast.

Life in this rich and rainy landscape has also led to a fascinating array of local adaptations by wildlife species whose ranges go far beyond the limits of the ecoregion. Wolves, thought to have evolved in open and relatively flat environments suited to hunting large herds of hoofed mammals, have mastered an existence

OPPOSITE: The trail of a trotting wolf strikes out across the beach on the west coast of Vancouver Island.

ABOVE: Bounded to the east by the Coast Range, British Columbia's central coastline is a maze of rainforest-cloaked islands. The marks of logging roads and cutblocks are apparent on several islands.

in these dense, rugged forests by making use of intertidal zones, at times exchanging running across large open tracts of land for swimming across channels of salt water between islands.

Although they were eliminated from the Olympic Peninsula early in the twentieth century, wolves never completely disappeared from British Columbia's central coast. On Vancouver Island, the largest island on the west coast of North America, wolves were also eliminated, or nearly so, in the middle of the twentieth century but then quickly reinhabited the island. Fortunately detailed and creative research into the lives of wolves in coastal British Columbia has teased out much about their life history and ecological significance in temperate rainforests. Here they eke out their living not only on deer and the occasional mountain goat like their interior cousins but also on salmon, seals, crabs, and other tasty treats from the sea. Their affinity to the water is so strong in this unusual landscape that photographer and regional conservationist Ian McAllister refers to them as sea wolves.

THE CENTRAL COAST

In the heart of the Great Bear Rainforest is Bella Bella, a small town on an island about one hundred miles north of the northern tip of Vancouver Island. My flight into Bella Bella helped me put the landscape into perspective. A vast network of forested islands spread out below, divided by emerald-green

Standing and fallen dead trees are an important part of the ecology of temperate rainforests, providing important habitat diversity for wildlife.

salt water. Glacier-capped peaks rose to the east. To the west, the islands tapered off into the vast Pacific Ocean. Out to sea fog and clouds lingered, indicating that even on that bright, early fall day, the mists and rain would never be far away. Departing from Vancouver's airport, we had traveled up the eastern shore to Port Hardy at the island's northern tip, and from there up the mainland coast to Bella Bella. The view changed noticeably as we traveled north. Vancouver Island and the mainland

Countless forested islands dot the central coast of British Columbia.

adjacent to it are defined by roads and large blocks of clear-cut forest. The farther north you go, this human footprint shrinks. As we dropped in altitude on the approach into Bella Bella, sporadic cut blocks still stood out but were now an exception on the dark green, forested islands that we passed over.

As striking and beautiful as this landscape is from the air, to really appreciate the Great Bear Rainforest you need to travel on water. During my trip to the area in the fall of 2011, I had the opportunity to explore the inlets and islands around Bella Bella with Douglas Brown, the field station manager for Raincoast Conservation Foundation at the time.

Since its inception in 1990, Raincoast has been among the most effective and innovative regional conservation organizations seeking to protect the Great Bear Rainforest, blending professional wildlife and ecology research efforts with well-designed and creative conservation initiatives. Using noninvasive techniques, research carried out through Raincoast has revealed a lot about the unique lives of the wolves of the rainforest and especially about their relationship with salmon. Raincoast's purchase of exclusive rights to run guided trophy hunts of wolves and bears in a large portion of the Great Bear has reduced the hunting pressure on these large carnivores. Their supporters, who care about the intrinsic value of the animals, pay to go on guided "hunts," demonstrating that many people place a higher value on the chance to see wolves and bears in the wild than on viewing a tanned pelt or stuffed trophy in the living room.

Doug had been the field station manager for five years when I met him, but his roots in the Great Bear go back much further. A member of the Heiltsuk First Nation, in whose territory much of Raincoast's efforts take place, Doug grew up in Bella Bella, learning to fish with his father, a commercial fisherman, and hunt and camp in the rainforest with his uncle. One day, while we were out on the water searching the coast for wolves, Doug pointed out the site of the historic village in which his family lived for several thousand years before being relocated to Bella Bella by the Canadian government in the late 1800s.

Raincoast's research has shown that the story of wolves in the Great Bear Rainforest is also, in large part, a story about salmon. The wolves eat salmon, a lot. This is not surprising really—just about everything in the Great Bear has some sort of connection with salmon. Terrestrial mammals from mice to grizzly bears feast on spawning salmon. Bald eagles and ravens flock to salmon streams to gorge on the fish, as do many other birds. I have watched unlikely species such as song sparrows and hermit thrushes flit out of the brush to peck at the rotting bodies of spawned-out salmon on the banks of a river. Birds and mammals fly off with or drag carcasses into the surrounding forests. One day, while searching under fallen logs in the rainforest several hundred yards from the banks of the Sol Duc River in the Olympic Mountains of Washington, I discovered the remains of salmon carcasses. Who dragged these large fish into the small space beneath the log is still a mystery to me, but it certainly helps explain how marine-derived nutrients brought by salmon far inland can be found in the vegetation and trees of the forests around salmon-bearing streams.

Animals will go to great lengths to take advantage of food abundance in order to avoid competition. Tiny birds, such as warblers and hummingbirds, will fly from the tropics to the Arctic to take advantage of the brief summer abundance of insects and flowers in a landscape free of most of their tropical neighbors with whom they winter. Some elk migrate from low elevations to subalpine meadows for the brief mountain summer, enjoying the lush summer grasses. Of these adamant travelers, Pacific Northwest salmon are among the most diligent. No matter how many times I have seen a two-foot-long, unreasonably large fish battling its way up a small stream where catching an eight-inch trout might otherwise seem like a coup, it never ceases to amaze me. Despite our natural sense of revulsion to the smell of rotting fish and the sight of scores of decomposing carcasses along the banks of (and floating in) the region's streams, these are the signs that all is well in the world during autumn across much of the Pacific Northwest.

THE WONDER
OF A WANDERING FISH

The life of a salmon sometimes begins only a few hundred yards from the sea, sometimes hundreds of miles from the ocean, but always in fresh water. Depending on the species, young salmon either head straight to the sea shortly after birth or linger for up to several years in fresh water before migrating to the ocean. Once in the Pacific, salmon born in the streams of the Northwest often range far and wide, up into the icy waters off the coast of Alaska and across the ocean to the coast of Asia. Hunters, salmon prey on smaller fish and invertebrates, and the ocean provides an abundance of food that no freshwater system can match. Oceans also have much lighter predation pressure on salmon than they find in freshwater systems.

Five of the region's seven species of salmon breed only once in their lifetime. Beckoned by the call of their homeland after one to five years at sea, they return to the mouth of the stream in which they were born. Apparently guided primarily by olfactory cues, most individual salmon travel up their natal freshwater stream toward the specific reach of water of their birth. About 10 percent of salmon err in their navigation and return to a stream other than the one they were born in, a characteristic of these fish which allows them to recolonize streams from which they had disappeared or pioneer new streams as environmental conditions give them access. This likely enabled a quick return of salmon to streams during the end of the last ice age as the glaciers melted and rivers returned.

While salmon are not the only anadromous fish to migrate from salt water to fresh water to breed in the Pacific Northwest, they are the most prolific. The freshwater streams here are home to seven species of anadromous salmon: chinook, coho, chum, pink, sockeye, steelhead trout, and sea-run cutthroat trout. In their lifetime steelhead and sea-run cutthroat trout may make multiple trips between the ocean and fresh water to breed. The remaining species make only one such trip and then die shortly after breeding. Each species prefers a different type of fresh water for spawning, a fact that allowed this guild of fish to spread across the region and occupy every ocean-going river system in the Northwest from northern California to Alaska.

The journey to a suitable location could be just a single day for some pink salmon that breed in small coastal creeks. For some chinook and sockeye, which make their way hundreds of miles inland, the journey could take weeks. Once in fresh water, no matter how long the journey, adult salmon (with the exception of steelhead and cutthroat) stop feeding. All their attention turns toward one purpose: reaching the waters in which they were born, in order to spawn. Females select a spawning location and use their tail to excavate a shallow depression in the gravel bottom of the stream, known as a redd, in which they deposit eggs. After a male fertilizes the eggs, the female buries the redd and guards the site. Once their mission is complete, death is not far behind for either sex, usually within a couple of weeks.

Humans aren't the only inquisitive terrestrial species to take an interest in the life history of salmon. Mammals and birds of all sorts bend their seasonal activities to salmon runs. The largest salmon runs of the year are generally in the fall. As the dry summer months give way to fall rains in September and October, the rise of Northwest streams draws salmon into the fresh water. This is perfect timing

ABOVE: A black bear carries his prize, a chum salmon, back to shore on a river in British Columbia's Great Bear Rainforest. Bears are one of the many species that take advantage of migrating salmon in the Pacific Northwest.

OPPOSITE ABOVE: Salmon, such as this decomposing chum salmon in a coastal stream in British Columbia, bring marine nutrients inland and provide an important food resource for a variety of animals as well as increasing the productivity of nearby plants and forests.

OPPOSITE BELOW: A pink salmon struggles in a shallow coastal stream, making it an easy target for fishing wolves or bears.

for many terrestrial fish-eating species looking to maximize their caloric intake before the rigors of winter set in. None of this, however, explains one burning question: exactly how does a wolf catch a fish?

Doug landed the boat as quietly as possible along the shore, around a bend from the stream we had been monitoring for a couple of weeks. It was not fully light yet, and we were hoping to get to the edge of a meadow a little way upstream from the creek's mouth before the ravens that had been lingering in the area woke up and drew attention to our approach. As Doug anchored the boat, I headed for the creek to scan for salmon. The tide was out, and the low water in the stream had left dozens of pink salmon splashing their way forward, in many places less than half submerged in the shallow water.

I picked my way upstream slowly, scanning the surroundings in the dim light. Ahead I could see the silver shapes of salmon lying in the grass along the bank of the stream, several dozen of them. As I approached, the distinctive signature of their slayer became clear: wolves. Each fish was either completely headless, or just the braincase had been carefully removed from the carcass. There was one large male pink salmon on the bank whose head was entirely intact. Odd, I thought, though in the frenzy of splashing fish and hunting companions I could imagine a wolf dropping the fish and heading back for another without consuming it. But then something happened that changed my opinion about why this fish still had its head. It flopped.

In the field of wildlife tracking, as in any scientific endeavor, additional data calls for a reassessment of hypotheses. How long had these fish been here? How long had that one intact salmon been there? I mentally sorted over the history of the tides as I scanned the salmon carcasses, looking for signs of deterioration since death. I reflected on my past experiences watching fish out of water and considered how long they might continue flopping before finally perishing. The fish were bright. A number of them were lying in the streambed in a location where the incoming tide would flood them shortly. A new story coalesced in my mind to explain all the evidence before me. A wolf had dropped that fish and left the area in a hurry because it had been startled by the approach of another type of piscivorous carnivore, namely Doug and me.

Doug approached behind me and surveyed the scene. He agreed with my assessment. We quickly moved toward the forest edge, one of us on either side of the stream to conceal ourselves and wait to see if the wolves might return.

This was not the first location where we had found signs of wolves feeding on salmon during the two weeks that I had been visiting. Several of the salmon-bearing streams we had been monitoring had evidence of wolves visiting close to where the streams empty into salt water. Raincoast researchers had been monitoring these streams for a number of years, and some of the observations and data that led to the first extensive scientific exploration of the role of salmon in the diets of coastal wolves had occurred on these streams. Doug had chuckled on one occasion when I had noted my excitement about getting to visit these locations after reading the dry, technical scientific journal articles that this research had produced. Doug, who had been involved in this research, noted that the Heiltsuk have always known about the relationship between wolves and salmon, but when a scientific research program documented it, many people felt like this relationship had just been discovered.

Chris Darimont, researcher and science director for Raincoast Conservation Foundation, agrees with Doug, believing that in many cases western science ends up confirming what local peoples have known, while bringing additional levels of detail to the conversation. To this end, Darimont, the lead author of numerous articles and reports on the ecological role and conservation of wolves in the rainforest, has illuminated a very different picture of the life histories of wolves than has been described in other parts of North America. He and his field crews diligently observed the fishing methods of wolves in the Great Bear Rainforest, measured the importance of fish in wolves' diets, and

A male (left) and female pink salmon showing the typical feeding pattern of wolves, with only the head or brain being consumed. Central coast, British Columbia.

looked at how this unusual food source influenced the carnivores' ecological relationships with other prey species. They discovered that salmon comprises 2 to 16 percent of the diet of wolves in the area but can make up a much higher percentage than this during the peak of the fall salmon runs when they can be the primary item in the diet.

Wolves select shallow streams where they can wade into the water easily. They generally face upstream, into the current, which allows them to approach upstream-swimming salmon from behind and carries their scent downstream and away from their quarry. They use both visual and auditory clues to find and home in on salmon swimming upstream. Once a salmon is located, wolves plunge their muzzles into the water to capture the fish with their teeth, then bring it to shore for consumption. In conditions where streams are very low, salmon are particularly vulnerable and in these circumstances wolves might approach from any direction. Only the head is typically consumed, though occasionally more of the carcass may be eaten if a wolf is scavenging along the shore rather than actually fishing or when salmon are very scarce, such as at the beginning or end of a run.

A collection of pink salmon fed on by wolves on the banks of a coastal stream in the Great Bear Rainforest.

Salmon provide several benefits for wolves. Foraging for salmon is far safer than attempting to subdue ungulates that can severely injure or kill wolves with their hooves. Because of the consolidated and predictable location of salmon, search time and energy devoted to finding food are also greatly reduced in comparison to searching for deer in the forest.

For wolves there is one potentially fatal drawback to feeding on salmon, however. Salmon host a parasite that canines (including domestic dogs) can contract from eating the fish. The parasite is usually fatal for canines if untreated. *Neorickettsia helminthoeca* is a bacteria-like organism that has a complex life cycle which includes passage through an intestinal fluke (*Nanophyetus salmincola*), an aquatic snail (*Oxytrema* species), various species of salmonids, and salmon-eating mammals including wolves. Snails, infected with the fluke, are consumed by salmon and trout. In the fish the fluke is benign, found most abundantly in the muscle tissue and kidneys of the fish. When the fish is consumed by a mammal, the larvae embed in the intestinal tract and the bacterium is released, causing symptoms including persistent diarrhea and vomiting. Cysts of the infected flukes are expelled in the feces of the infected mammal. Foraging snails pick up these cysts and the life cycle begins again.

By only consuming the head of salmon, wolves reduce chance of exposure to this parasite, as the head is less likely to be infected than the rest of the fish. The brain is also a highly nutritious portion of the fish, so that in the presence of such

an abundance of food, wolves may be eating only the preferred portion of the fish. Bears may also occasionally select the head of salmon, but they typically consume the eggs and portions of the body as well. Systematic documentation of wolves eating salmon is recent, but the first regional documentation dates back to the late 1800s when a die-off of wolves in the Oregon Coast Range was suggested to be a result of "salmon poisoning," perhaps infection with the bacterium *Neorickettsia helminthoeca*.

Research has documented salmon in the diets of both interior wolves and coastal wolves in southeast Alaska. Salmon was a larger part of the diet of coastal wolves, however. Historically, as in southeast Alaska, salmon were ubiquitous and abundant across the Pacific Northwest, including the entire Columbia River and Snake River watersheds as far as hundreds of miles from the coast. The range of salmon in the Columbia Basin has been greatly reduced, and the abundance of salmon in remaining stocks is much lower than it was in the past. It is likely that wolves across the region once made use of this valuable food resource, much as they do in areas where both wolves and salmon continue to coexist elsewhere.

Wandering along the banks of streams such as Bear Valley Creek in the Salmon River Mountains of central Idaho, one can still watch three-foot-long chinook salmon making their way to spawning grounds in the cool, shallow creek. If fish populations increase enough to catch the attention of the reestablished wolf populations here, they may be added to the menu once again.

Shortly after Doug and I settled into our respective locations along the stream, the low, eerie sounds of multiple wolves howling from just to our north drifted through the mist. The response of a single wolf soon followed to our south. A few minutes later, the single wolf from the south emerged from the forest slightly upstream from us. Undetected by the wolf, we watched as it moved to the banks of the stream and howled. It entered the water and moved upstream out of my view. From his vantage point, Doug watched the wolf move up the stream, then notice a lone pink salmon in shallow water. It quickly descended upon the fish, plucked it from the water, and carried it to the shore where it consumed the head. It returned to the location where it had howled from. We both watched as a raven swooped down at it playfully from its perch on a nearby tree. The wolf appeared to take little notice of the bird but soon retreated back into the forest.

Besides salmon, coastal wolf populations make extensive use of other unique marine food sources. Wolves hunt and forage in the intertidal zone, where prey items include harbor seal, river otter, mink, and such marine invertebrates as crabs, clams, barnacles, and even squid. In one wolf population, living on a collection of islands relatively distant from the mainland of British Columbia, such marine foods accounted for 52 percent of the wolves' diets, with salmon making up an additional 16 percent.

However, even in this unique landscape where the sea and the land come together, wolves still usually depend heavily on ungulates. Along the central and southern coast of British Columbia, black-tailed deer are ubiquitous. Various studies have shown that deer comprise between 32 and 82 percent of wolves' diets. Roosevelt elk have a much more limited distribution and population size, though in one study from Vancouver Island they were up to 28 percent of wolves' diets.

A raven swoops down at a wolf on the edge of a salmon bearing stream in the Great Bear Rainforest.

Wolves evolved as predators of hoofed mammals. What is it about the Great Bear Rainforest that has led to wolves, at times, getting the majority of their caloric needs from other sources? In the islands off the central coast of British Columbia, as deer abundance decreases, the importance of salmon and marine mammals in wolves' diet increases. Similarly, on Vancouver Island, where forestry practices have decreased deer abundance in coastal locations, wolves relied more heavily on a secondary prey source, beavers in this case. A similar pattern might be expected in heavily managed forests with low deer abundance in western Washington and Oregon.

While clearcuts in temperate rainforests maintain their carrying capacities for deer and elk immediately after harvest, within several decades the dense, closed-canopy forest produces little forage for ungulates. These have been dubbed "ungulate barrens" in southeast Alaska and British Columbia. Not surprisingly, industrial-scale logging not only affects deer abundance but also influences secondary prey selection by wolves. In logged areas in southeast Alaska, river otter consumption increased, while in unlogged forests salmon consumption was higher.

In clearcut forests, salmon populations often decline because of the destruction of spawning habitat. Most of the low-elevation and coastal forests in western Washington and Oregon are heavily managed for timber harvest with cut-cycles that do not allow the forests to mature to the stage when they begin to increase in productivity as wildlife habitat. The largest remaining runs of salmon in Oregon and Washington are in coastal creeks and rivers, and if wolves manage to establish themselves in these areas, we might see similar patterns to those in coastal forests to the north.

The diet of wolves has attracted the attention of another intelligent coastal forager. All across the world, wherever wolves and ravens coexist, people have noted the close relationship between these two species. Pacific Northwest First Nations in particular recognize the unique relationship between wolves and ravens. The common raven (*Corvus corax*) is the largest member of the family Corvidae , which comprises a collection of intelligent and adaptable birds found across most of the globe. Ravens are adept scavengers, which is the most important connection

Three ravens surround a wolf-killed salmon carcass.

between the bird and wolves. Corvids, opportunistic by nature, are attracted to carcasses in general. But the relationship between ravens and wolves may be more intricate than ravens just showing up at wolf-killed carcasses.

In coastal portions of the Pacific Northwest, ravens are abundant and their affinity for wolves is hard to miss. An evening of hunting by wolves can produce days of feasting for these large, smart black birds. After Doug and I had encountered wolves wrapping up a fishing session, we returned regularly to the same location over the next week. Seven days later we still found about a dozen ravens feeding on the remains of the same pink salmon the wolves had pulled from the stream. In my own experience across the Pacific Northwest, I have found ravens in close association with wolves, often following the wolves as they travel through the landscape and perching in the trees in locations where wolves are resting or feeding, and twice nesting close to a wolf den site. While ravens may see wolves as the potential provider of food, the two species might appreciate each other's playful nature as well. On several occasions while exploring the coast of British Columbia, I observed ravens swooping down out of trees close above a wolf's head in an apparently playful gesture.

What ravens get out of this relationship is clear. One study in the Yukon reported that the impact of raven consumption of wolf kills might double the kill rate necessary for a pair of wolves. For this reason, despite their generally tolerant and sometimes even playful relationship with ravens, wolves occasionally kill ravens at carcasses.

If it is clear why ravens benefit from the relationship, the value for wolves, if any, is a bit less clear. Adept scavengers, ravens fly over the landscape in search for carcasses or injured animals. Both wolves and humans, also intelligent and curious scavengers, attend to the wanderings of ravens. The collection of multiple ravens in a particular area always draws me in to see if they are at a carcass feeding. While in the field searching for wolves, I often pay careful attention to the movements and calls of ravens and jays (another corvid). Their calls and behavior in the presence of wolves, bears, and other large animals are distinctive and have often alerted me to the presence of animals before I detected them myself. Similarly, I have watched other wildlife attend to the movements and calls (or sudden cessation of calls) of corvids and other birds. From the raven's perspective, attracting such attention from a human or a wolf might lead to the dismemberment of an intact carcass. When a large animal dies, ravens are unable to access most of the carcass because their bills are not designed to pierce the thick skin of larger mammals. Drawing in wolves or humans may be a self-serving act to get one of these other species to reveal the goodies contained beneath the thick hide. From a wolf's perspective, paying attention to ravens might lead to an easy meal.

VANCOUVER ISLAND

"Now!" Steve Lawson commanded as he revved the two large outboard motors on his hand-built, aluminum skiff and pulled up to the sea-spattered rocks on the shore of one of the outer islands in Clayoquot Sound. My feet dangled off the bow of the boat. On his command I pushed off the boat and jumped to the rocks, fifty pounds of photography equipment on my back. Instantly Steve had both engines in reverse, and the boat was twenty feet away from the rocks and swiveling around to face the ocean swell about to roll in. I scrambled for higher ground to avoid the surf that was about to crash onto my landing zone.

Steve and Susanne Lawson, veterans of decades of battles to protect Vancouver Island's ancient forests and salmon runs from industrial logging and, more recently, marine fish farms, know the nooks and crannies of the island's west coast as do few others. Between homeschooling their children and battling timber interests, fish farmers, and province bureaucrats, they have explored just about every inch of the beaches and rocky shores of Clayoquot Sound's numerous islands from their home near the town of Tofino. Steve had suggested that the stretch of beach to the north of where he dropped me off might be a hotspot for wolf activity. In fact, the previous day we had spotted one leisurely walking the shore in the midafternoon.

Steve and Susanne pulled out and headed north to a promising fishing location, leaving me to explore, hoping that the swells didn't get any larger, which might make it impossible for them to come back and pick me up. I cut across the rocks to the sandy beach and quickly picked up the tracks of the wolf from the day before. I backtracked it out onto the beach. Multiple other wolf trails crisscrossed the shore and were joined by the trail of a black bear and several black-tailed deer. Remote and almost impossible to land a boat on, the beach held no visible trace of anything human. I wandered in and out of the dense rainforest that loomed above the shore, following tracks coming and going from the beach.

Clayoquot Sound sits on the west coast of Vancouver Island. Here rainforest-clad mountain slopes and an archipelago mark the end of land, giving way to the cold, wild waters of the Pacific. With a human population of only four and a half million, mostly located in the southern portion of the island, Vancouver is primarily defined by forest. A rugged mountain range, rising to 7219 feet at its highest point, runs along the center of the island, with numerous alpine glaciers at its higher elevations. The island's west coast, exposed to the full brunt of Pacific storms, is rain-soaked and dominated by dark rainforests. The eastern portion of the island, while by no means dry, sits in the rain shadow of the island's tallest peaks, and here slightly more moderate climatic conditions prevail. Farther east, the mainland east of Vancouver Island is often referred to as British Columbia's Sunshine Coast.

The story of wolves on Vancouver Island has some distinctly different twists from that of the nearby central coast of the mainland. Several reasons for this became readily apparent the sunny July weekend I drove off a ferry and headed northwest across the island toward the west coast to meet up with Steve and Susanne. Hoards of traffic, full campgrounds, and the bustle of Tofino made it quite clear that this was a landscape both appreciated and well used by Canadians. During the two-and-a-half-hour drive to Tofino, a second major difference between the island and the central coast was also apparent. Making sure not to veer into oncoming traffic so as to avoid being smashed to bits by an oncoming RV or log truck, I gazed up at mountainside after mountainside of clearcut forests in various states of regeneration. While large tracts of the island are still inaccessible and relatively undisturbed by resource extraction and recreation activities, about three-quarters of the island's forests have been roaded and logged. Logging started on the island in the late eighteenth century, but industrial scale clearcut logging really began in earnest in the 1960s and peaked in the 1980s. As easily available timber has disappeared, cut rates have declined, but timber harvesting remains the most important industry on the island, including continued cutting of old growth forest stands.

The entrance at the base of this large western red-cedar descends into a wolf den which had been excavated under the tree. Large red-cedar trees such as this one are a common feature of wolf dens in coastal British Columbia.

In many ways, besides just geographic location, Vancouver Island represents a middle ground between the more remote coast to the north and the more densely populated, roaded, and developed coastline and coastal mountains to the south in Washington and Oregon. The plight and progress of wolves on the island reflects parts of the story from both the north and the south. Wolves in the Great Bear survived the species' widespread extirpation from most of the rest of the region. South of Vancouver Island, wolves in western Washington and Oregon disappeared by the mid 1900s and have now been absent for many decades. On Vancouver Island wolves appear to have naturally reestablished themselves after a relatively brief extirpation.

Wolves on Vancouver Island were once considered a distinct subspecies (*Canis lupus crassodon*). If the original population were still present today, it is unclear whether it would be considered a distinct subspecies based on modern criteria. Regardless, British Columbia's wolf eradication programs destroyed, either completely or nearly so, the island's wolf population by the 1960s. The island's current wolf population consists either entirely or primarily of descendants of animals that dispersed to Vancouver Island from the mainland shortly after the cessation of predator control activities. These wolves moved across the archipelago in Desolation Sound, which separates the mainland from Vancouver Island, and established themselves in the northeastern portion of the island. From here, wolves expanded south and west, eventually reoccupying the entire island in locations that are still habitable for the species.

A wolf pup splashes through a shallow coastal stream. Den and rendezvous sites in the rainforest are often associated with salmon-bearing streams such as this one.

On Vancouver Island, wolf population density estimates have been as high as anywhere in North America. One study estimated one wolf per 2.3 square miles, if just the lower elevations within wolves' ranges were surveyed, or one wolf per 4.6 square miles if entire watersheds, including high elevations, were used. Along with high population densities, one study found that pack home range sizes, documented at about 43 square miles, were small relative to the average home range size for packs in other locations in North America. Both the high density of wolves on the north end of Vancouver Island and their small home range were attributed to high prey (black-tailed deer) density at the time of the research. As with wolves elsewhere, substantial dispersal distances have been documented on Vancouver Island. One radio-collared male traveled from the northeast of Vancouver Island nearly to the southern tip over a period of seven and a half months, wandering at least 215 miles from the site where it had been captured and collared.

With wolves occupying heavily logged landscapes as well as areas that consist primarily of old growth forests, researchers have detected some variation in the ways wolves use these habitats. Studies of den sites on Vancouver Island, however, showed that in fragmented habitats, wolves still selected den sites in mature forest stands. In both pristine and fragmented habitats, rendezvous sites are most often located at small open meadows, either in mature forests or along the coast.

BEACHCOMBING WOLVES
AND BLACK BEARS

The interaction of predator and prey is one of nature's most basic ecological relationships. A fundamental challenge for many prey animals involves getting the resources they need to live while avoiding detection by predators. Conversely, predators must have methods for finding these prey animals that are working so hard to avoid being found. Ranging widely is in large part a method that wolves use to find prey dispersed across the landscape. This life-or-death game of hide and seek

A wolf travels along the water's edge on a small island in Clayoquot Sound.

can often be detected in the tracks of predators and their prey: bobcats and rabbits, mountain lions and bighorn sheep, mink and muskrats. It's actually a fairly simple pattern: the prey species must leave its secure habitat to feed, usually by a speedy and direct path, while the predator chooses a hunting route perpendicular to that of the prey, along the edge that the prey must cross to get from security to food. The muskrat must leave the security of deep water, where it can outmaneuver a mink, to feed on the banks of a stream. The mink travels the edge of the stream, increasing the odds of catching a foraging muskrat on the banks and placing itself between the muskrat and the water. Rabbits leave the safety of their thicket to feed on grass in the meadow, and the bobcat stalks along the thicket's edge, hoping to catch the rabbit out in the open.

On several islands along the outer coast of Clayoquot Sound, the two most common mammal tracks I discovered on the beaches were those of wolves and bears. Wolves here eat a wide variety of intertidal creatures that they discover during their shoreline sojourns. Terrestrial mammals, including black bears, also use

the intertidal zone for foraging and in so doing put themselves at risk of predation from wolves.

While wolf trails typically follow the shore for long distances, I found that bear trails more often emerged from the dense forest and poked around on the beach, often among the seaweed and flotsam at the high-tide line, before returning back into the forest. Early one morning I watched a black bear emerge from the rainforest and forage in the washed-up seaweed. Dissection of several bear scats in the area later led me to believe that it had been feeding on sand fleas, which were abundant in the seaweed. After the bear departed, I walked along the beach inspecting numerous bear tracks, probably all from this one bear, going back and forth from the forest to the ocean's edge. I also found numerous tracks of wolves traveling along the shoreline up and down the beach. It dawned on me that I had seen this pattern of movement in two species before—between predators and prey.

Just before dawn this black bear came out of the forest to feed on sand fleas along the ocean's edge. It lingered long enough for a photograph in the dim early morning light before returning to the forest. Away from the security of the trees, black bears are more vulnerable to wolves. Clayoquot Sound, British Columbia.

Bears are primarily forest creatures: their ability to climb trees provides an important refuge from the only carnivores that predate on adult black bears in the Pacific Northwest—wolves and grizzly bears, neither of which is an effective tree climber. Being caught out in the open without access to the security of trees can be a deadly mistake for a black bear in wolf country. In his book *The Last Wild Wolves*, Ian McAllister recounts observing a black bear that was swimming across a channel and accidentally landed close to a pack of wolves on the shore in the Great Bear Rainforest. The bear bolted across the shore for the safety of the trees but didn't make it and was summarily dismembered and consumed by the wolves. Biologist Bob Hansen told me that on the west coast of Vancouver Island, deer populations are sparse and black bears relatively abundant, perhaps in higher total numbers than deer. He hypothesizes that black bear, and specifically bear cubs, might be an important supplementary food source for wolves in some coastal locations with low deer populations.

As in other parts of the region where the two species overlap, wolf interactions with grizzly bears in the rainforest are a bit less predictable than those with black bears. While grizzly bears usually command the best salmon-fishing spots on coastal streams in British Columbia, packs of wolves have been observed displacing grizzly bears at such locations. In interactions between competing carnivores, weight usually defines interactions, with the larger animal dominating. However, the social nature of wolves appears to thwart this trend.

SWIMMING

Wolves living on islands, hunting bears on the beach, and eating fish, crabs, squid, and seals clearly raises the question of how they get around in a landscape defined as much by water as by land.

Several days before Steve Lawson's two daring feats of boating (he successfully retrieved me from my exploration of the remote beach on Clayoquot Sound), we were traveling through more protected waters when Steve noticed a slender wolf moving along the shore of another, much smaller island, not more than twenty-five acres in size. Something about how it was moving made Steve believe it might be about to enter the water. We idled offshore and watched. Eventually the wolf came down to the water's edge, where it paused to sniff the wind and survey its surroundings before entering the sound. With just its head above water, it made its way steadily across the channel toward the shore of the adjacent island about two hundred yards away, appearing much as a pet dog might if swimming determinedly toward a stick its owner had thrown into the water. The water of the small channel reflected the dark green forests and clear skies, and the wolf left a small wake that rippled across the still water. It approached the far shore and micro-navigated toward an area with an easy exit from the water. It emerged dripping wet and quickly shook from head to tail, sending water flying in all directions. Then it set out along the shore, foraging along the water's edge on the dropping tide. Now its businesslike demeanor clearly distinguished its wild nature from the tendencies of its domestic cousins. For this wolf, the swim was a pragmatic behavior designed to get it to the next foraging location.

How often do rainforest wolves take to the water? Our observation, while not common, was certainly not unheard of. Many fishermen on the coast have stories about coming across wolves, deer, and bear swimming between islands in the Great Bear and around Vancouver Island. Observers in the Great Bear have documented wolves making their way to islands seven and a half miles from the next closest piece of land!

LESSONS
FROM THE NORTH

As humans have persecuted wolves across most of the Pacific Northwest, the rugged coastline, vast forests, towering peaks, and limited human population of British Columbia's coastal rainforests have acted as a refuge for wolves and many other species over the past several centuries. Now they are also a source for the reexpansion of wolves into landscapes to the south.

What does the future hold? These lightly inhabited islands and mainland shores of the Great Bear Rainforest give us a glimpse of what coastal Oregon and Washington were once like. Farther south on the coast, Vancouver Island has a history of exploitation that more closely resembles that of the altered coastal landscapes in Oregon and Washington. How wolves have adapted to the roads and logged landscapes on the island, as well as to its more heavily populated and recreationally attractive coastline, foreshadow what may unfold farther south if wolves are able to establish coastal populations again in Washington, Oregon, and northern California.

No place on this planet is immune to the changes that humans have wrought, however. While they have survived several centuries of persecution by western civilization, the wolves, bears, salmon, and trees of the Great Bear face mounting and diverse threats. Along with continued industrial logging operations across the area, salmon farming operations are undermining the vitality of many of the Great Bear's salmon runs, one of the ecological foundations of this unique ecosystem. Marine fish farms, set up in inlets along the coast of Vancouver Island and much of the mainland coast, have brought increases in sea lice that infect wild salmon as well as escapes of captive Atlantic salmon. In many rivers where wild salmon must pass fish farms to reach their natal freshwater streams, salmon runs have crashed despite having high-quality spawning habitat.

A further threat to the region comes in the form of a proposed oil pipeline, which would carry crude oil from Alberta, east over the Rockies and Coast Range, and to the port of Kitimat, in the northern portion of the Great Bear Rainforest. From the port, oil would be loaded onto tankers that would then have to navigate miles of narrow inlets before making it to the open ocean. The waters of this proposed route are at least as complicated as those of Prince William Sound, where the Exxon Valdez disaster occurred, and the risk and consequences of a large-scale spill would be on a similar scale.

Salmon, a cultural icon of the Pacific Northwest for both First Nations and modern western culture, have been a vital resource for humans coastally and far into the interior of the region for uncountable generations. Just as for wolves along the central coast of British Columbia, much of the annual cycle of human activity in that area revolved around being prepared for fall salmon runs. The decline of salmon following European-American colonization of the region has led to ecological and cultural transformations of many landscapes in the Northwest. In places such as coastal British Columbia, salmon continue to be abundant; these fish remain economically and nutritionally vital to people and wildlife. Humans, bears, eagles, and an array of other scavengers and predators congregate at salmon-bearing streams during spawning runs to take advantage of this amazing gift from the sea.

ABOVE: Emerging from the water, the wolf shook off before continuing its travels along the shoreline.

OPPOSITE: From a small boat we watched this wolf make its way between two islands in Clayoquot Sound on an overcast summer day.

Salmon and other marine foods are probably the most unique aspect of wolves' diets here in the Pacific Northwest. As we are drawn together around a shared resource, the role of salmon in the lives of both wolves and humans points out yet again the intertwined nature of our two species.

Two fish counters from the Heiltsuk Fisheries Program walking along a stream counting the number of salmon that have returned.

On one of my first days out in the Great Bear, Doug Brown landed the boat on the shore of an island close to a salmon-bearing stream. The tide was out, and to one side of the stream's outlet, making an arc through the barnacle-covered cobbles of the intertidal zone, was an ancient, seaweed-covered rock wall. Doug explained that this was a traditional fish trap constructed many generations ago by the Heiltsuk people. Before running up the stream, salmon will mass in the waters around the mouth, waiting for the fall rains to raise the stream level and give them access to their spawning ground. As the tide drops, some fish would become trapped behind the wall as the waterline falls below its top. The Heiltsuk could then easily retrieve the salmon at low tide. Not surprisingly, wolves also recognized the potential of these human-created structures for fish collecting, and made use of them.

This rock wall built in the intertidal zone adjacent to a salmon-bearing stream is a traditional fish trap, constructed by the Heiltsuk people many generations ago. Wolves have been documented to use these traps to collect salmon.

This is a terminal fishery, or a place for catching salmon where they return to fresh water as opposed to out at sea. The Heiltsuk and other coastal First Nations people, along with the wolves and bears of the Great Bear, made use of terminal fisheries for thousands of years before European-American settlement. In the middle of the twentieth century, amidst concerns over declining salmon runs, British Columbia's provincial fisheries ministry decided that these traps and terminal fisheries were contributing to the declines, and they broke most of the traditional traps by opening up a gap in the wall at low points in them. More recently, in yet another ironic return of western civilization to traditional ecological understandings of resource management, fisheries science has once again begun to explore terminal fisheries as a possibly more effective way to conserve salmon runs, as catches can be more directly linked to rates of return to specific streams. For their part, wolves were never disillusioned with terminal fisheries. While humans have switched to other methods of fishing, wolves continue to take advantage of these traps to collect fish wherever they can.

SIX

WHERE DID THEY COME FROM AND WHERE ARE THEY GOING?

Т HE BURKE MUSEUM of natural history and culture, on the University of Washington campus in Seattle, houses a replica of a cavern created in a basalt formation in eastern Washington by lava flowing around the carcass of a prehistoric rhinoceros that once roamed the Pacific Northwest. Visitors can also view the fossil remains of a giant sloth, affectionately called the Sea-Tac Sloth (discovered during construction work at Sea-Tac International Airport between Seattle and Tacoma). There are also remains and reproductions of saber-toothed cats and a mammoth, creatures that wandered the Pacific Northwest along with wolves within the past fifteen thousand years. During a visit to the exhibit, I pondered the thought of running into a feline with six-inch-long fangs while rock climbing in the basalt cliffs along the Columbia River, or coming around the corner on a hiking trail in the Cascades to find a group of fifteen-foot-tall hairy elephants blocking the way.

As amusing or ridiculous as this might sound, the reality is that the first solid archeological evidence of humans in the Pacific Northwest overlaps with the last evidence of many of these awesome creatures by several thousand years, both linked to the end of the last period of continental glaciers in our region. People right here were probably having such encounters while hunting, foraging, fishing, and traveling across the Northwest. Shaped by massive geologic and climatic events, as well as by the range expansions and contractions of various plant and animal species, this region has been an extraordinarily dynamic place for thousands of years.

The landscape encountered by prehistoric Northwesterners was far different from the one their descendants were living in when European and European-American explorers first made contact with them. Similarly that landscape— dominated by unbroken coastal rainforests, rivers flush with salmon, widespread interior grasslands inhabited by large herds of grazing animals, and arid interior mountain ranges covered with forests sculpted by generations of anthropogenic burning practices—is now, in many instances, almost as distant a memory as the topography scoured by the retreating ice thousands of years earlier.

Throughout the Pleistocene epoch, *Canis lupus* moved into and out of the northern part of the region as the continental ice sheet and alpine glaciers waxed and waned. With the advance of the continental ice sheets, wolves retreated south (and possibly west to ice-free coastal locations). At times the wolf has been a single element of a broad guild of large carnivores that included numerous larger and more powerful species, such as social hunting lions similar to the modern African lion, and larger species of canines such as the dire wolf. Since the Pleistocene extinctions that swept across North America at the end of the last ice age, *Canis lupus* occupied the role of apex carnivore, possibly in habitats where it was previously excluded by recently extinct larger carnivores, as are wolves in modern times in the Russian Far East by Amur tigers, and African wild dogs are by African lions in sub-Saharan Africa. From the end of the last period of glaciation until the region's eighteenth-century European-American colonization, the species could probably be found in just about every landscape in the Pacific Northwest.

Following the arrival of European Americans, the distribution of wolves once again contracted, this time to remote locations where they were safe from human persecution. Wolf populations were decimated, with a steep acceleration of this process starting in the middle of the nineteenth century. By the middle of the twentieth century, wolves in the Northwest were limited to the far northern portions of the region, in the inaccessible British Columbia coast and interior mountains. In recent decades, however, with the rebound of ungulate populations (which had also been decimated by European-American settlement) and shifting human behavior toward wolves, their distribution has been expanding again. In the modern Pacific Northwest, wolves are finding a niche in landscapes that have changed significantly in many places from the ones that their ancestors once roamed across.

Wolves have been wanderers since the origin of the species. The modern gray wolf (*Canis lupus*) evolved from *Canis lepophagus*, a now extinct, smaller, coyote-like species that originated here in North America but traveled to Eurasia via the Bering land bridge. The first fossil records of *C. lupus* are thus from Eurasia, but it then retraced its ancestral steps, expanding its range into North America. In fact, fossil and genetic records suggest multiple migrations of *C. lupus* into North America from Eurasia over tens of thousands of years, possibly associated with advances and retreats of continental ice sheets. This ancient record of large-scale dispersal and population reestablishment is intriguing in light of the species' modern history. As wolves reestablish themselves in the Pacific Northwest, we are experiencing just the latest installment of a process that has occurred several times before as this species responds to changing environmental conditions and shifting ecological relationships with other large carnivores on the landscape—in this instance, humans.

Determining where wolves will end up and assessing the significance of their historical distribution for their future distribution is complicated. Humans, a behaviorally flexible species, are by far wolves' single largest modern predator.

PREVIOUS: *A wolf emerges from a dense western red-cedar forest.*

How our species treats wolves regionally is subject to individual personalities, the shifting winds of politics, and the predilections of a rural population with a strong streak of anti-government-regulation sentiment. Further, the Pacific Northwest is now undergoing significant climate shifts that affect many of the region's plant and animal communities.

Western ecological thinking has begun to view ecosystems as dynamic mosaics in a state of constant flux. In contrast, wildland communities used to be seen as relatively static, a perspective based on the misconception that if modern humans haven't directly changed a landscape, such as through logging or mining, then its current state is likely how it has always been. But ecosystems are constantly in motion, and wildlife distributions bend with them, which is important context for understanding the history and future of wolves in Pacific Northwest wildlands.

WHERE WERE THEY?

Historically around the globe, wolves occupied just about every landscape containing large hoofed mammals north of about 20 degrees latitude. Here in the Pacific Northwest, about fifteen thousand years ago nearly all of British Columbia and the northern portions of Washington, Idaho, and Montana were essentially wiped clean by the advancing ice of the Wisconsin glaciation period, with the possible exception of some ice-free locations along the coast. During this time wolves likely roamed south of the ice in present-day southern Washington and Oregon, as well as in an ice-free refuge in the interior of Alaska and the Yukon. The ice began to retreat as early as thirteen thousand years ago and had completely retreated from the region by nine thousand years ago, with the exception of the alpine glaciers which still exist. With the retreat of ice, wolves and their prey recolonized all of the Pacific Northwest. Hoofed mammals adapted to every ecosystem in the region included pronghorn (*Antilocarpa americana*), elk (*Cervus elaphus*), and bison (*Bison bison*) on the grasslands of the Columbia Plateau; bighorn sheep (*Ovis canadensis*) and mountain goats (*Oreamnos americanus*) in canyon and mountain regions; and deer and moose in forested regions. Beavers, an important secondary prey for many populations of wolves, were also abundant across most of the region. Coastal landscapes provided marine food for a significant part of wolf diets there. Finally, the entire region's abundant salmon runs provided yet another source of nutrition lacking elsewhere in North America that increased the productivity of Pacific Northwest landscapes for wolves.

After the end of the last period of continental glaciation, wolves ranged across the entire Northwest. By the middle of the twentieth century, the species range had contracted to the northern portions of British Columbia and the Canadian Rockies. Following their legal protection, wolf populations began expanding again.

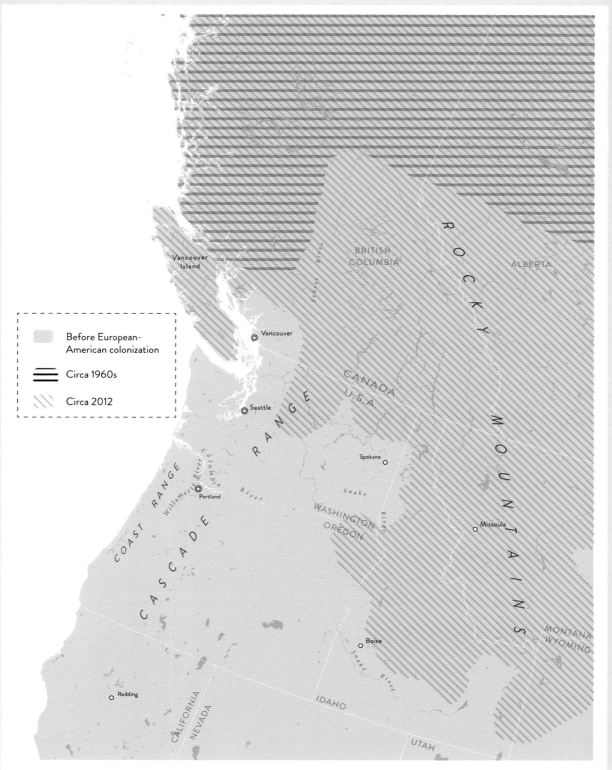

WOLF DISTRIBUTION IN THE PACIFIC NORTHWEST

European-American settlement of the Northwest heralded large-scale changes in ways that modern residents of the region may find hard to fathom. The earliest of these changes were poorly documented. Historical records predating intensive fur trapping, habitat conversion, introductions of livestock, agriculture, and associated invasive plant species, predator control efforts, and diseases are spotty and inconsistent. These gaps in the historical record, in turn, might have resulted in disproportionate reporting of wolves in some areas, while they could have been completely exterminated from others without any documentation.

For instance, several sources, relying primarily on museum collections and bounty records, claim that wolves were probably most abundant west of the Cascades. Another source states that wolves were apparently historically absent from the Columbia Plateau altogether. With elk, deer, bison, and pronghorn antelope rambling around the grasslands of the Columbia Basin back in the day, it is highly unlikely that wolves bypassed such targets in a landscape they are well adapted to hunt in.

Similarly two academic reports, also using a limited data set for determining the modern distribution of mountain lions, another wide-ranging and low-density carnivore, omit all of southeastern Oregon, while I found definitive sign of this species in every mountain range I visited during fieldwork in the area. If the data and method used here can make such omissions for a species' current range, we should probably be even more suspicious of inferences made about the historic range of a species. It might be more accurate to consider any records of wolves as an artifact of their persistence in these areas long enough for their demise to be noted. Or these records may in fact offer a snapshot in time of the range or abundance of wolves here on the path to their ultimate disappearance from most of the region.

Also obscuring our understanding of historical wolf habitat are limitations in records passed down by early settlers who mistakenly believed the landscape they encountered was in an original state. In fact, the ecological conditions in many parts of the region when they arrived had been shaped by the land-use practices of indigenous peoples, including landscape-burning and plant-cultivation techniques. Further, the end of indigenous practices in the face of European-American colonization in turn initiated noticeable ecological changes as well. In some instances, these changes even preceded the arrival of the colonists themselves in the form of disease, the introduction of horses, and the onset of the fur trade.

Information for determining the historic range, abundance, and habitat use of wolves regionally comes from a variety of sources, including fossil records, direct observations documented in explorer and settler journals, trapping records, bounty records, and the traditional ecological knowledge preserved in the stories, first-person accounts, and ritual life of First Nations people. We can also use what we know about modern habitat preferences and other factors that influence wolf distribution today to help piece together an understanding of historic wolf presence in our region.

As hard as it might be to imagine as you sip a latte while window shopping the art galleries and tourist shops of Friday Harbor on San Juan Island, or ride a ferry past the vacation homes dotting the shores of the islands today, wolves roamed this Puget Sound archipelago into the 1800s, documented by European explorers and settlers. An archeological dig at a historic indigenous village site on San Juan Island recovered the remains of wolves along with the bones of elk, moose, black bears, mountain goats, and Dall sheep, a subarctic species. While some of these remains could have been carried to the island by prehistoric hunters, in sum they hint at a collection of fauna quite different from that of modern times. Similar remains have been found in archeological sites all around the region; the area's earliest fossilized wolf remains, collected in Oregon, were well over 300,000 years old.

From at least the end of the last period of glaciation, about ten thousand years ago, wolf populations have been influenced by humans here in the Pacific Northwest. First Nations groups both competed with wolves for prey and may have suppressed populations of elk, deer, and moose in some locations. Conversely, indigenous groups also actively modified some Northwest landscapes to increase the abundance of ungulates, which could have had a positive effect on wolf populations. Landscape-burning practices maintained open forests and grasslands, supporting high abundances of hoofed mammals which indigenous groups and wolves alike depended on for food.

Human impacts on wolf populations came on an entirely different scale with the advent of the fur trade and European-American colonization. With the establishment of fur trading posts in the Northwest by the Hudson's Bay Company in 1821 and the subsequent establishment of bounties on wolves in 1843, we have both our first systematic documentation of wolves in the region and systematic efforts to destroy them. Prior to 1821, significant fur trapping had already been occurring, at first focusing on sea otters until their near extinction in the early 1800s, then shifting to beaver and other terrestrial furbearers.

Records from four Hudson's Bay Company trading posts located within the boundaries of the current state of Washington and one located close to present-day Kamloops in southern British Columbia document the extraction of 15,995 wolf pelts from the Pacific Northwest between 1821 and 1859, an average of 421 a year for nearly four decades. Contrary to assertions that wolves were historically more abundant west of the Cascades, the largest number of wolf pelts were traded overwhelmingly at the interior trading posts at a ratio of twenty-three to one, although the interior posts likely collected pelts from a larger territory than that of the coastal locations to the west. Nevertheless, these numbers clearly demonstrate the former abundance of wolves in interior landscapes. The single-year wolf pelt record from the Nez Perce trading post of about eleven hundred in 1847 represents two hundred more wolves than the estimated population in the entire state of Idaho in 2009 when the species was removed from the USFWS endangered species list for the first time. Indeed, this may have represented a significant portion of the entire population in that part of the Northwest as the harvest level dropped precipitously in the following year and to nothing by 1856. Random reports from various locations in Washington and Oregon documented continuing populations of wolves through the early 1900s. The last consistent reports of wolves in Oregon came during the 1930s on the west slope of the Cascades and from the Olympic Mountains around the same time in Washington.

Wolf populations probably hit their lowest levels and most contracted range between the 1950s and 1970s. At this point, wolves were functionally, if not completely, extinct from all of Oregon, Washington, Idaho, Montana, and the southern portions of British Columbia including Vancouver Island.

WHERE ARE
THEY COMING FROM?

After more than a century of persecution in the United States and Canada, wolves began reexpanding their range naturally following the abatement of control efforts in British Columbia and Alberta and the official protection of wolves under the Endangered Species Act in the United States portion of the region in the 1970s. The continued growth and expansion of wolves into vacant habitat in southern British Columbia, Washington, Oregon, and eventually northern California has been fed by wolves from three sources. Wolf populations along the central coast of British Columbia began expanding south along the coast. In the northern interior of British Columbia and Alberta, wolves began expanding south through the mountains. In the 1990s, translocations would further expand their numbers.

All wolves in the Pacific Northwest today originate from populations from the British Columbia coast and the northern interior of British Columbia and Alberta.

DNA studies of the rainforest wolves in coastal British Columbia have shown some distinct genetic signatures compared to wolves from the interior of the region or elsewhere in North America. After predator control efforts in the province abated, wolves from the central coast began successfully dispersing south, reoccupying Vancouver Island. In fact, the distinct genetic signature of this wolf population has made it south all the way into the Washington Cascades.

At the time of greatest contraction in their range, wolves were absent from the Rocky Mountains far north into Canada. They were first documented breeding again in Jasper and Banff national parks in the early 1980s. Farther south, just across the border in Montana, wolves were documented breeding on the west slope of the Rockies in the valley of the North Fork of the Flathead River in Glacier

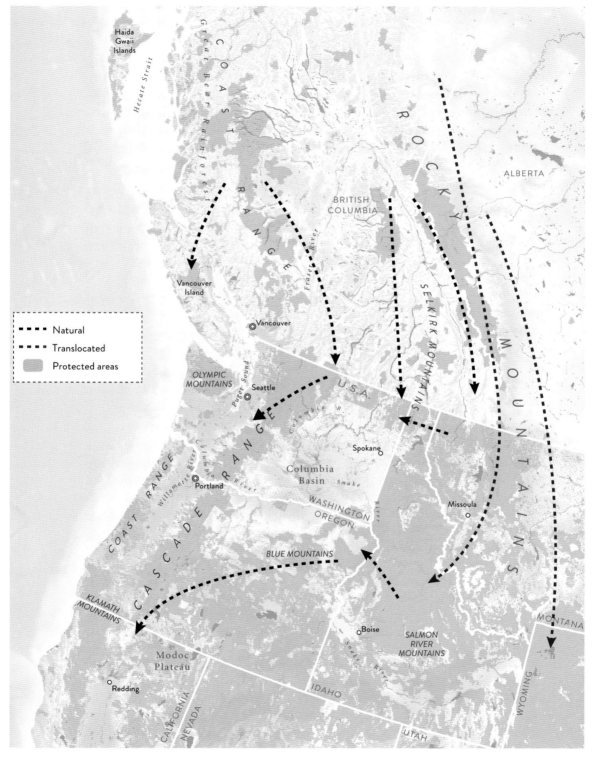

WOLF POPULATION SOURCES AND DISPERSAL IN THE PACIFIC NORTHWEST

National Park by the mid 1980s. This naturally expanding population has grown over time, and dispersing wolves from the Canadian and U.S. Rockies have moved south in the Rockies and west through the Idaho panhandle and into the Selkirk Mountains and Columbia Highlands of Washington.

Wolves translocated to central Idaho and Yellowstone National Park represent a third population source for the Pacific Northwest. As part of the wolf recovery program in the northern Rockies, the USFWS translocated sixty-six wolves from Canada to central Idaho and Yellowstone National Park between 1995 and 1996. Wolves were trapped close to Hinton, Alberta, and Fort Saint John, British Columbia, about 621 miles and 932 miles north, respectively, from where they were released. Wolves from these areas represented the closest population source from an ecologically similar environment where the USFWS could obtain animals for the reintroduction. The wolves in northeastern Oregon are related to the central Idaho population which are at least primarily descendants of this reintroduction effort.

The classification of wolf subspecies in North America has been a topic of much interest and controversy over the years. It has been a source of intense debate within the biology and conservation worlds and a source of political tension between advocates and groups opposed to wolf reintroductions in the Rocky Mountains. Groups opposed to the introductions claimed that the U.S. government introduced the wrong subspecies into central Idaho. However, there has been general agreement that older classification schemes which listed as many as twenty-four subspecies of wolves in North America (including six in the Pacific Northwest) were inaccurate. Based on contemporary subspecies classification, the Pacific Northwest is currently home to only two distinct subspecies: *Canis lupus nubilus* and *C. lupus occidentalis*, of only five subspecies across all of North America. The range of *C. lupus nubilus* is believed to include most of western North America including the entire Pacific Northwest coast east to the Great Lakes, while *C. lupus occidentalis* occupies much of Alaska and the interior of western Canada, dipping south into the northern Rockies of the United States. While the extirpation of wolves from large portions of North America certainly wiped out a lot of genetic diversity, the reorganization of subspecies classifications reflects a better understanding of the actual genetic similarities and differences among wolves in North America both historically and currently. The translocation of wolves from central British Columbia and Alberta to Idaho and Wyoming was essentially speeding up a process of dispersal that was already occurring naturally.

Besides putting into question historic subspecies classification schemes, research has also begun to explore how both ecological setting and geographic proximity affect wolf population genetics in North America. Wolves would be

The most recent wolf taxonomy identifies two subspecies in the Pacific Northwest.

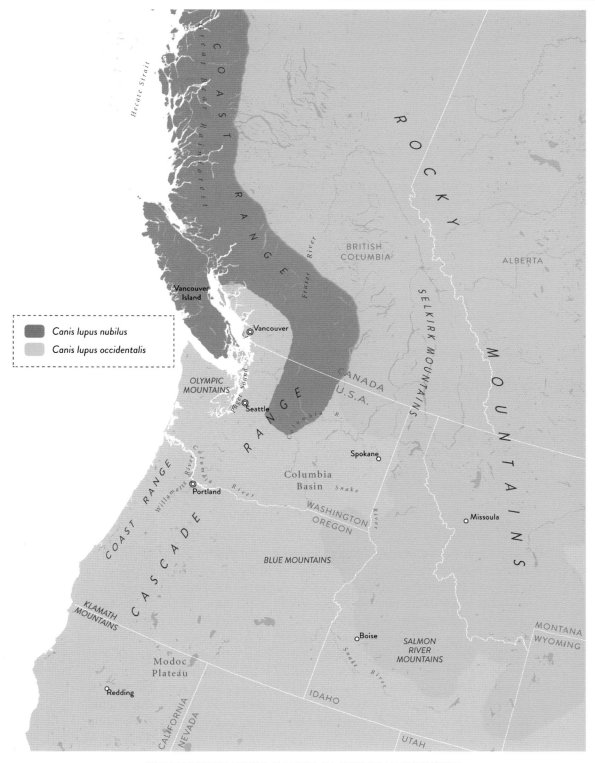

Legend:
- Canis lupus nubilus
- Canis lupus occidentalis

Hecate Strait

COAST RANGE

Great Bear Rainforest

Fraser River

ROCKY MOUNTAINS

BRITISH COLUMBIA

ALBERTA

Vancouver Island

Vancouver

SELKIRK MOUNTAINS

CANADA U.S.A.

OLYMPIC MOUNTAINS

Puget Sound

Seattle

Columbia R.

RANGE

Spokane

Columbia Basin

Snake

COAST RANGE

Willamette River

Columbia River

Portland

WASHINGTON OREGON

Missoula

CASCADE

BLUE MOUNTAINS

Snake River

KLAMATH MOUNTAINS

Modoc Plateau

Boise

SALMON RIVER MOUNTAINS

MONTANA WYOMING

Redding

CALIFORNIA NEVADA

IDAHO

UTAH

WOLF SUBSPECIES IN THE PACIFIC NORTHWEST

expected to have a higher degree of genetic likeness across long distances than animals with smaller home ranges and shorter dispersal distances. However, research now suggests a stronger correlation between ecological habitats and wolf population genetics than between absolute proximity and genetics. Wolves with more similar genetics tended to occupy ecologically similar locations irrespective of the distance between populations, with ecological similarity in location accounting for 70 percent of genetic variation and the greatest similarity found in wolves from similar ecological zones. Conversely, distance alone accounted for only 31 percent of variation, with more distant populations being genetically more dissimilar regardless of habitat. Researchers hypothesized that because wolves learn to hunt certain prey in certain conditions they may be predisposed to disperse into similar areas rather than to just the closest available habitat.

Researchers who looked specifically at the genetic distinctness of British Columbia's coastal wolves came to similar conclusions. Here, the Coast Range appears to act as a strong ecological barrier for dispersal into and out of this coastal region, with far more genetic interchange up and down the coast. This dynamic also raises the question of what the historic genetics of wolves along the Washington and Oregon Coast were like. Based on habitat characteristics, one might reasonably hypothesize that these wolves shared more genetic characteristics with existing northern coastal populations than with Rocky Mountain and interior wolves. In Oregon, the only source population is the Rocky Mountains, so it is likely that wolves with this genetic lineage will end up populating the west slope of the Cascades and Coast Range. In Washington State, the Lookout pack's genetic similarities to British Columbia's coastal wolf population have demonstrated the connectivity between coastal British Columbia wolf populations and those of the North Cascades, contributing to a more diverse assemblage of wolf genetics here in the future.

WHERE ARE THEY GOING?

Liberated from both Pleistocene competition and more recent systematic hunting, trapping, and poisoning campaigns, wolves are once again on the move in the Pacific Northwest. But where will they end up? What makes a good home for wolves in the region? Washington's Diamond pack, whose territory straddles the Washington-Idaho border and the crest of the Selkirk Mountains, illustrates the complexity of this question. This pack inhabits a landscape fragmented by an extensive road network, a checkerboard of land ownership, and a mosaic of clearcuts interspersed with dense stands of regenerating second and third growth mixed conifer forests. Moose populations have been increasing steadily for decades here as have white-tailed deer (*Odocoileus virginianus*), likely taking advantage of increased forage associated with the conversion of closed canopy forests to open shrub fields in the wake of clearcuts. About three hours from Spokane, the northeast corner of

Washington, which the Diamond pack calls home, is among the most sparsely populated parts of the state.

My attempts to locate the pack during the winter of 2010 to 2011 were thwarted despite a ridiculous abundance of logging roads dissecting the landscape. Getting around in the winter in the pack's home range is painfully difficult. This is in part due to the lack of clear maps demarcating roads and the fact that many roads are gated and locked. In three days of skiing snowed-in roads and bushwacking on skis through dense, dog hair conifer stands on a sickening, icy raincrust, I never got closer than two miles from a recently known location for the pack.

Far from a pristine landscape, this part of the Selkirks is very much a working forest. During the 2010 breeding season, the pack's den was located close to an area that had been logged during the previous winter. Meanwhile, just a little way to the north, the higher elevations and unroaded, uncut forests of the Salmo-Priest Wilderness appear rarely to be used by wolves. What is it about the rolling, roaded, and cut landscape that the Diamond pack finds attractive? The answers to this question illuminate a great deal about what makes good wolf habitat in the modern world, an important step in predicting the future distribution of the species in the Pacific Northwest.

A number of models have been created to predict how wolves might reestablish themselves in Washington, Oregon, and California. These studies, based on evaluations of occupied and unoccupied landscapes in the northern Rockies, attempted to assess the relative value of particular landscapes in four general categories: prey density, prey accessibility, connection to other quality habitat, and security from human-caused mortality. Variables included density of ungulate species, road density, human population density, topography, livestock abundance, and land ownership. In general landscapes with high ungulate population densities, low human influences (such as remote roadless areas), moderate topography, high public lands ownership, and low livestock numbers would be ideal habitat as these areas would provide high levels of accessible food where wolves would be secure from human-caused mortality. Much of the Selkirk Mountains, with abundant moose, elk, and deer populations, low human population density, and minimal livestock grazing, rank high in such models. In contrast, most of Puget Sound, despite a variety of food sources, ranks low because of very high human population densities and associated infrastructure.

Taking the criteria one by one, prey density denotes sufficient wild ungulates or other primary food sources to sustain wolves throughout the year. The rebound of wild ungulate populations across the region over the past century has once again created abundant food sources for wolves. However, in the Pacific Northwest wolves may persist even in the face of low numbers of ungulates where salmon and marine food sources heavily subsidize their diets.

Prey accessibility refers to the relative ability of wolves to capture prey in a particular landscape. Studies of existing wolf populations in the Rocky Mountains and the Columbia Mountains of British Columbia have shown their light use of steep and high-elevation terrain and greater use of valley bottom or more gentle topography. This may reflect increased prey densities in these areas but also the wolves' increased ability to access prey here. Mountain goats, caribou, and other hoofed mammals use steep and high-elevation terrain as a deterrent to wolf predation and thus avoid being a substantial part of their diet in most our region. Because of this, models assessing the potential value of landscapes for wolves place higher value on lower-elevation habitat and landscapes with more gentle topography.

Another important factor affecting the relative likelihood of wolves setting up shop in a particular place is their ability to get there at all. Imagine you're a wolf living in the Cascades around Mount Saint Helens in the southern portion of the Washington Cascades and you want to emigrate to the Olympic Peninsula because you have heard about abundant deer, elk, and salmon, and loads of high-quality yet unoccupied den site locations. What could be better from a wolf's perspective? Maybe a little less rain and some more sunshine in the winter, but most of us here in the Northwest are willing to make this tradeoff for other quality-of-life perks that come from living here. Unfortunately for this wolf, getting to all that abundant and accessible prey would require running a gauntlet of highway crossings, farm fields, suburban developments, and heavily roaded timberland. Just Interstate 5 alone greatly reduces the chances of a large carnivore dispersing from the Cascades to the Coast Range.

This lack of connection between various landscapes that could otherwise support wolves decreases their overall value for the animal. Conversely, while the steep slopes and relatively low ungulate densities of the North Cascades are perhaps not ideal for wolves in terms of those values, this area's location adjacent to existing wolf populations—connecting those landscapes to areas with better prey density and accessibility—makes it very valuable wolf habitat. Connectivity between parts of a landscape is also important in established populations, allowing for intermixing of genetics and repopulation of an area that has had a high degree of mortality from disease or humans.

Within the geographic range of any animal population there are often sources and sinks. Population sources are locations where reproduction outpaces mortality, and thus the number of animals increases. Sources generally feed dispersing young animals into adjacent areas. Population sinks are locations where mortality rates exceed birth rates, and therefore the number of animals in the area shrinks if there is no augmentation from source areas. For wolves, national parks and inaccessible wilderness areas are often sources since here wolves are protected from human persecution, while agricultural and rural roaded landscapes tend to be sinks as wolves are exposed to a much higher level of legal and illegal human persecution. Wolves have an exceptional propensity for dispersal across long distances.

Combined with a high reproductive rate for a carnivore of their size, they are able to send numerous animals into adjacent or distant landscapes.

Conversely, population reductions through human interventions, such as wolf hunts in British Columbia, Idaho, and Montana, likely reduce the rate of dispersers into unoccupied areas: fewer potential dispersers remain, and local breeding opportunities increase with the loss of resident breeding animals. That wolves have been legally hunted continuously across mainland British Columbia has likely slowed the rate of population growth there and the rate of dispersal south.

Humans are the single largest source of mortality for wolves in the Pacific Northwest and adjacent regions through legal and illegal hunting and lethal control of wolves that prey on livestock. Landscapes where humans have minimal access, where wolves are less visible to humans, and where wolves are less likely to interact with livestock have a decreased likelihood of human-caused mortality and increased security for wolves. Various habitat models have used forest cover, human population density, road density, level of road use, percentage of public versus private land ownership, and amount of livestock grazing activity as measures of the relative security of landscapes for wolves.

Taking another look at the Diamond pack's territory in the Selkirk Mountains, we can see how, from a wolf's perspective, they have done very well for themselves. While the area is heavily roaded, much of the road system is gated and inaccessible to humans in motor vehicles. The low human population here further reduces potential conflicts with wolves. When compared with the higher elevations and contiguous forests of the Salmo-Priest Wilderness, the area the Diamond pack occupies likely has more moose, who use clearcut and regenerating forests extensively. Since moose are the only hoofed mammal whose primary winter range overlaps extensively with the pack's, their abundance is likely vital. The topography where the Diamond pack lives is noticeably gentler than that of the Salmo-Priest Wilderness to the north, another feature predicted by the models we're looking at here. Finally the pack's territory is well connected to adjacent areas of the Idaho panhandle, northwest Montana, and southern British Columbia with previously existing wolf populations.

While the Diamond pack has carved out a territory in Washington which provides everything they need to flourish, the complex topography and wide array of human land use patterns across the region have created large variations in the quality of habitat for wolves here. This will likely lead to a patchwork wolf distribution in the Pacific Northwest. The absence of any one of the four primary habitat criteria—prey density and accessibility, landscape connectivity, and security from humans—would make the establishment or long-term survival of a wolf population more challenging. Wolves currently recolonizing northeastern Oregon have faced challenges related to high livestock levels and large tracts of private lands. In other areas, wolves' needs are pitted against each other. The Lookout pack's territory exemplifies this. Surrounded by huge tracts of steep-sloped, high-elevation

ABOVE: The view from a wolf rendezvous site in the Great Bear Rainforest. Coastal landscapes offer a diversity of food resources for wolves.

OPPOSITE ABOVE: This large boulder hidden in dense forest in the Selkirks creates excellent refuge characteristics for a wolf den. Adjacent to the boulder, note the well-worn ground used extensively by pups and adults during the denning season.

OPPOSITE BELOW: A large chamber excavated under the boulder provides protection for pups when adults are away hunting.

roadless wilderness areas that provide excellent security from humans, the wolves instead spend much of the year at lower elevations in close proximity to the local human population in areas used for livestock. Why? These lower-elevation, gentle slopes are the winter and spring range for the deer population, these wolves' primary food source. As predicted from models, the pack has faced a high level of poaching.

In the southern portions of the Northwest—the southern Oregon Cascades and Modoc Plateau, which spans the Oregon-California border—is the largest patch of high-quality habitat in southern Oregon and northern California. Additional habitat lies along the entire crest of the Oregon Cascades and in the mountains of the northeast corner of the state where wolf reestablishment has already begun. The Oregon Coast Range has been described as marginal habitat, with the likelihood of high levels of human-caused mortality because of high road densities. However, similar to areas in the Washington Selkirk Mountains, many roads in the Oregon Coast Range are closed to the public, so they don't pose the same threat to wolves as open road systems do.

High densities of deer and elk, moderate topography, and large quantities of public forest lands around Mount Saint Helens in the southern Washington Cascades make this area high-quality wolf habitat.

In Washington, different models have come up with a variety of scenarios for quality wolf habitat, depending on what data sources they used and which parameters they measured. The lower-elevation rolling terrain, several large roadless wilderness areas, abundant deer and elk herds, and relatively small human population in the central and southern Washington Cascades were consistently identified as excellent habitat. Farther north in the Cascades, higher elevations, steeper topography, and smaller ungulate populations make this landscape of lower value but still habitable by wolves. The forested mountains in the northeastern corner of

the state rank highly, as does the core of the Olympic Peninsula.

The open landscapes of the Columbia Basin in Washington are uniformly seen as low-quality habitat given the low population densities of wild ungulates, high visibility of wolves there, large amounts of agriculture, and access by humans. The high deserts of eastern Oregon are a little more favorable but also rank as low-quality habitat.

WOLVES ON THE MOVE: PATTERNS OF DISPERSAL

Once young wolves reach sexual maturity, they have a few options for establishing a territory and breeding. These include attempting to carve off a portion of their parents' territory (usually including some adjacent areas as well), usurping their parents' position within the existing pack and territory, or most commonly, dispersing to unoccupied territory. Dispersal can be a short distance to a location immediately adjacent to their natal territory, or it can be several hundred miles away. The reports of a young radio-collared male which dispersed from its natal pack in northeastern Oregon in 2011 captured national, and even some international, news headlines. OR-7 crossed the state to the southwest before heading further south to California where it lingered for several months before returning to southwest Oregon and then back again to California, a trip of well over 700 miles. Such traveling exploits are not atypical for dispersing wolves.

Researchers have found that even if adjacent locations have a suitable prey base and habitat features, some young wolves will disperse long distances before settling. In other instances, wolves on the periphery of existing wolf range will disperse back into more densely populated areas where competition would be greater. Dispersal is of vital importance for expanding wolf populations, as it is the primary source for wolves to occupy new areas. Between 10 and 40 percent of a wolf population disperses from its natal territory annually.

For over a decade I have been teaching people how to navigate cross-country through the North Cascades as an instructor for Outward Bound. A rugged range of steep-sided, densely forested ridges, rocky inaccessible ridgelines, and U-shaped glacially carved valleys often make travel and navigation daunting for novices and experienced mountaineers alike. As students become more proficient in broad-scale route selection and micro-navigation, I take a back seat, letting groups read and interpret maps and make choices about specific travel routes on the ground. Following one such student group illuminated for me how the landscape drives the travel choices of both humans and wolves as they make their way across unfamiliar ground.

One August day, I was traveling with a student group over a trailless alpine pass in the Lake Chelan–Sawtooth Wilderness, west of Rennie Peak. As we came up from the north side of the ridge, the landscape was dominated by sheer rock faces, steep talus fields, and fields of snow, protected in the shade of the mountain peaks rising above them. The pass we were heading to was the only route out of the valley that wouldn't involve scaling the rock faces. From our camp, it stood out as a low notch between peaks that rose on either side of it. Leading up to it was an obvious route through a finger of trees and then open talus fields which the students quickly identified on the maps and then in the landscape the evening before from camp.

Topping out on the ridge in the small pass, we found a landscape that changed abruptly. On the south-facing slopes below, dense forests beckoned. On the rolling slopes above, hoary marmots whistled and wildflowers bloomed in the meadows. The students checked their maps and set out down into the dense subalpine forest, heading toward a trail in the valley bottom and camp beyond. Walking twenty yards behind the group, I watched them select the best route through the tangle of vegetation, avoiding fallen trees and the steepest slopes, using deer trails for brief sections, and moving toward more open areas.

After years of traveling through such terrain myself and watching hundreds of students learn to move through it as well, I could predict their choices which were punctuated by mistakes where the navigators reach impenetrable underbrush or an unexpected cliff band in the forest. At this point a brief backtrack and recalibration was required. Several hours into the bushwhack the students came to a small but fast-moving stream where two forks converged. This was the second time I had a group of students travel this general route and come to precisely this point to cross the stream. No trails or signposts lead to the spot, each group had to make dozens of route-finding choices from the starting point at the pass to arrive at this location, and yet somehow, improbably, they had. Humans, like wolves and all animals, have their preferences for travel and route selection on a landscape, defined by our unique mode of travel and desire to cover the most terrain in the safest and most efficient way possible.

On a broader scale, if wolves were able to create a transportation map of the Pacific Northwest, it would look quite a bit different from the road maps we are accustomed to looking at for this purpose. Instead of highlighting interstate freeways and secondary highways, it might highlight diffused corridors of the landscape, defined by geography, habitat, and human uses which represent likely travel and dispersal routes. The routes would connect larger landscapes that provide habitat for large numbers of resident wolves, much like the metropolitan areas of Puget Sound or the Willamette Valley would do on a human road map. Overlaying a human map with one made by wolves would produce some interesting areas of overlap and highlight areas of conflicting use—such as an interstate running east-west through an area identified by wolves as an important travel corridor running north-south.

Wolves don't randomly disperse across landscapes. Safe and efficient travel routes are important for every species of wildlife, but especially so for such a wide-ranging animal. The relative ease or difficulty which a landscape poses for travel is referred to as its permeability. On both a broad landscape level and in a micro-habitat level, certain terrain features are more or less inviting to wolves for travel. Some landscape features may be complete obstacles to dispersal, while others deter some but not all travel. These obstacles may be natural, such as the Strait of Juan De Fuca between Vancouver Island and the Olympic Peninsula, or created by humans, such as the corridor of development along Interstate 5 which runs through the Puget Trough and Willamette Valley in Washington and Oregon respectively. The same is true of humans. The small pass in the mountains my students navigated over would likely deter many humans, and if it hadn't existed at all, my group would have been forced to turn back as well, the surrounding mountains not being passable for us.

For millennia the Pacific Northwest has been a landscape with significant geographic features defining routes of dispersal for wolves and other wildlife. The genetics of wolves along the British Columbia coast suggest limited dispersal across the rugged Coast Range, while dispersal up and down the coast has been more common. In the western portion of the region, large bodies of water have acted as barriers to travel as well. The arrival of European Americans introduced several new variables with significant impacts on the permeability of the landscape for wolves and many other species of wide-ranging wildlife. Large blocks of urban areas, destruction of prey species, and multi-lane highways all currently provide additional challenges for dispersing wolves.

Large sections of the Olympic Peninsula remain similar to when wolves were last present there in the dense forests of the Olympic Mountains, protected in Olympic National Park. This location might seem to be some of the best habitat for wolves. However, because of a lack of adjacent source populations combined with large, possibly insurmountable, barriers (both natural and created by humans) to permeability, this area has not been the first location to be recolonized by wolves. In fact, despite excellent habitat on the Olympic Peninsula, it might be close to impossible for wolves to disperse there naturally.

OPPOSITE: Clearcuts, while providing excellent browse for deer and elk shortly after being cut, show a significant decline in carrying capacity for hoofed mammals for decades as the young forest canopy closes. Roads created to access timber allow for more human access and in turn a higher rate of wolf mortality. Western Cascades, Washington.

ABOVE: Wolves, such as this one in northwestern Montana, often use lightly traveled roads as efficient travel routes despite the increased likelihood of human encounters.

The howl of a wolf and many images that wolves conjure in people evoke deep wilderness and a spirit of wildness. Interestingly enough, though, the apparent association of wolves with wild, rugged, roadless country has little to do with their preferences for such landscapes and more to do with persecution by humans. What we see as either a noble or evil association between wolves and wild or primeval landscapes is actually a reflection of our own species' ability to shape the behavior of wildlife. Certainly a wolf running along the forested ridgelines of the North Cascades, stopping to howl in the moonlight, creates a more archetypal image of wildness than that of an animal slinking into town to make away with scraps of bones from behind the local butcher shop. Yet both of these images reflect the habitat and behavior of wolves here in the Pacific Northwest.

In parts of British Columbia wolves actually select landscapes with roads and associated logged forests as these modifications increase moose populations. Furthermore, in places where humans show benign or positive behavior toward wolves—such as in national parks or in cultures generally accepting toward wolves, as in parts of Europe where wolf populations have been expanding—wolves carry out their lives in close association with humans. In Yellowstone, wolves hunt, travel, socialize, and raise their young literally before the eyes of thousands of interested humans.

The impacts of roads on wildlife distribution and ecological functioning of landscapes has evolved into its own branch of ecology—road ecology. Large highways, such as the Trans-Canada Highway and U.S. interstates 90 and 84 in our region, have been demonstrated to significantly limit or alter the movement and dispersal patterns of numerous large carnivores. Roads, through motor vehicle collisions, are also a source of direct mortality for wolves and other wildlife. While both of these impacts of roads are important for wolf habitat and distribution in the Northwest, more compelling is the association between roads and other sources of human-caused wolf mortality such as hunting, trapping, and poisoning.

Lightly traveled roads are often used by wolves for travel and hunting despite the increased likelihood of encounters with people. Similarly, human use of landscapes changes and generally increases with the addition of roads, further increasing the likelihood of interactions between the two species. Numerous studies have clearly identified that wolf mortality increases with proximity to roads. On the British Columbia coast, where most human traffic is by boat, shoreline acts similarly to roads in this regard.

During my own fieldwork I have found roads to be some of the best places to detect wolves. In mountainous terrain they tend to use the valley bottoms disproportionately to their abundance in the landscape while humans often route roads through these same valley bottoms. Similarly humans select routes for roads in other parts of mountain landscapes that overlap with wolf travel, such as along gentle ridgelines and through mountain passes. Wolves select efficient travel routes. In roadless areas, wolves use game trails on ridgelines and valley bottoms as well as maintained or historic human trails. In terrain with significant winter snow accumulation, ease of travel on compacted roadways and snowmobile trails attracts use by wolves and other wildlife. For wolves such pathways allow more efficient access to broader areas, thereby increasing the potential for detecting game. Such roadways will also, in some instances, attract prey species for the same reason, thereby further increasing the likelihood of their detection by wolves.

Yet not all wolves respond in the same way to roads. Wolf populations that have lived in association with higher levels of human activity learn to make use of human-altered landscape features and to minimize the risks associated with them. In more remote settings, activities associated with roads can be very disturbing to wolves and can cause them to significantly alter their landscape use in order to avoid this disturbance.

Ever since the initiation of fur trapping and bounties by Europeans and European Americans, human land use and behavior have been the defining factor for wolf distribution in the Northwest. Adult wolves have no true natural predators save humans. Human persecution has been repeatedly identified as the largest source of mortality in wolf populations, even in legally protected populations, and has been identified as a limiting factor in their reestablishment in several areas. It is almost certain that in parts of the region, simply on the basis of habitat conditions and prey base, wolves could exist, but human choices—either within official management objectives or illegal persecution—will prevent wolves from establishing breeding populations. However, Oregon and Washington's management plans allow some level of assistance for the dispersal of wolves into areas where landscape fragmentation may cause reestablishment of a population to take a very long time.

The distribution of wolves in the Pacific Northwest will likely grow and change in fits and starts over the coming decades as new packs establish themselves and dispersing wolves show up a hundred or more miles from any other known populations and attempt to carve out a breeding territory. While the time of hairy elephants and six-hundred-pound cats may have passed forever here, the next chapter in the story of wolves is now unfolding. How wolves will adapt to, navigate, and make use of both the ancient and novel landscapes that make up the modern Northwest from the rainy shores of the Pacific to the arid deserts of the northern Great Basin will be a tale filled with predictable themes and unexpected discoveries for both the wolves and humans.

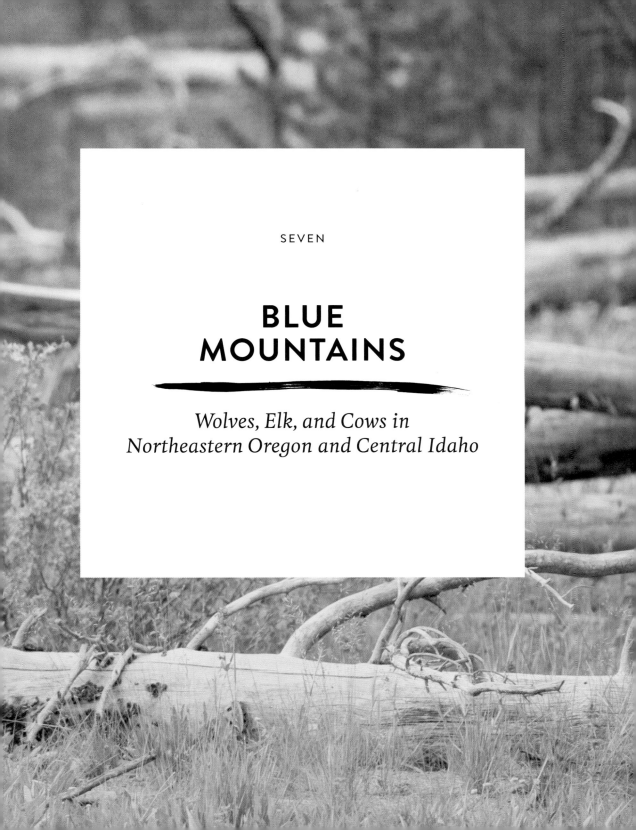

SEVEN

BLUE MOUNTAINS

*Wolves, Elk, and Cows in
Northeastern Oregon and Central Idaho*

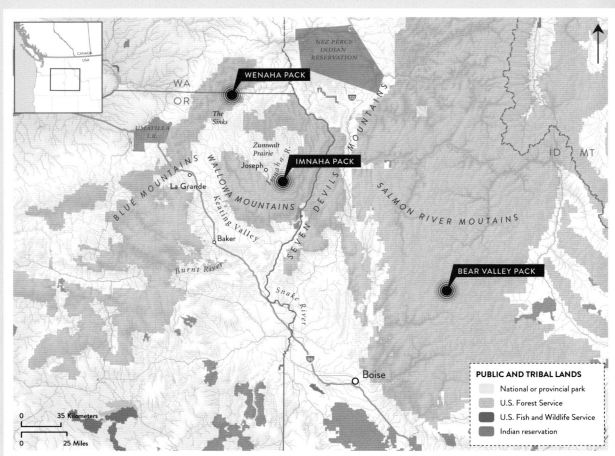

CANADA
USA

WA
OR

NEZ PERCE
INDIAN
RESERVATION

WENAHA PACK

The Sinks

UMATILLA
I.R.

Zumwalt
Prairie

IMNAHA PACK

Joseph

BLUE MOUNTAINS

WALLOWA MOUNTAINS

La Grande

Keating Valley

Baker

Burnt River

Snake River

SEVEN DEVILS MOUNTAINS

SALMON RIVER MOUTAINS

ID   MT

BEAR VALLEY PACK

Boise

PUBLIC AND TRIBAL LANDS

National or provincial park
U.S. Forest Service
U.S. Fish and Wildlife Service
Indian reservation

0   35 Kilometers
0   25 Miles

BLUE MOUNTAINS AND SALMON RIVER MOUNTAINS OF
NORTHEASTERN OREGON AND CENTRAL IDAHO

ALONG THE NORTHERN border of Oregon and Idaho, the Snake River runs through the bottom of a deep cleft in the mountains—Hells Canyon. Here, before it stagnates behind a series of dams, the largest tributary of the Columbia River is remote and wild. As one ascends into the mountains on either side of the river, arid grassland and shrub-steppe conditions at the bottom of the canyon give way to pine and fir forests. To the east, the Seven Devils Mountains span a scantly traveled section of Idaho backcountry. To the west, the Imnaha River has cut its own deep wandering canyon through the mountains as it makes its way to the Snake River. Slightly farther west, the high peaks of the Wallowa range in the Eagle Cap Wilderness, crown jewel of the Blue Mountains, rise steeply above the Zumwalt Prairie and the towns of Joseph and Enterprise. Ecologically and geologically, the Blue Mountains of northeastern Oregon and southeastern Washington comprise the westernmost extension of the famous, continent-spanning Rocky Mountains.

It is here in the Blue Mountains that wolves first returned to the state of Oregon. The first wolf documented to have dispersed to northeastern Oregon from Idaho did so 1999, a female that was ultimately captured and returned to Idaho. Between 2000 and 2007 three more wolves were found dead in northeastern Oregon, two of them shot, all dispersers from Idaho.

During the summer of 2006, biologists in the mountains northeast of Boise, Idaho, trapped and radio collared a young female wolf in the Timberline pack as part of ongoing efforts to monitor Idaho's wolf population. The wolf was given the name B300. A year and a half later B300 began to wander from her natal territory. She traveled more than one hundred miles northeast into Oregon, swimming across the Snake River in the process, and set up shop on the western edge of the Eagle Cap Wilderness and the Zumwalt Prairie. In the summer of 2009, she was retrapped along with an adult male, also a migrant from Idaho according to genetic analysis. Researchers also documented pups at that time, making this the first documentation of breeding wolves in Oregon since their extirpation in the 1940s. The pack was named the Imnaha pack, after the Imnaha River, an important landscape feature in their territory.

B300 and her mate chose a predictably desirable locale for wolves—home to a large herd of elk, ample mule deer, the occasional white-tailed deer, and the

---

PREVIOUS: *The canyons of the Imnaha River country are home to Oregon's Imnaha pack, named after the river.*

recent addition of moose (who themselves only in the past couple decades swam across the Snake River to become established in the Blue Mountains). Based on unpublished data collected by the Oregon Department of Fish and Wildlife (ODFW) from radio-collared animals in the pack, the annual home range size of the pack has been as large as 870 square miles. There was one hiccup in this wolf paradise, however: livestock ranching is widespread across the region. In the summer most of the rolling mountain terrain managed by the USFS is leased, for a nominal fee, to ranchers to run their cattle or sheep.

My first visit to this corner of the northwest was several years before the Imnaha pack's founders wandered in. Having driven along the interstate between Boise,

Oregon's first radio-collared wolf, B300, just after her release.

Idaho, and Portland, Oregon, several times, I had seen the snow-capped peaks of the Eagle Cap Wilderness off in the distance. I was keen to explore the area and jumped on the opportunity to join a buddy of mine, Al Thieme, on an elk-hunting scouting trip there. We were not the only folks who have taken an interest in the area in recent years. Abundant elk and open rolling landscapes make this one of the most exciting elk-hunting locations in the state. The wild rivers, amazing powder snow, and remote wilderness lakes also attract boaters, fishermen, backcountry skiers, snowmobilers, backpackers, and horseback riders.

As with many places in the rural interior of the Pacific Northwest, livelihoods of the residents of Wallowa county traditionally have revolved around livestock and timber. Socially, the transition to a service- and tourism-based economy has not been a pleasant one for many. An influx of urban refugee settlers, looking for a beautiful escape from the city, and outdoor recreationist visitors mixes poorly with the preexisting descendants of an earlier influx of European Americans.

On my second trip through the town of Joseph, I went searching for wolves. When I stopped into the local sporting goods shop to buy some topographic maps, I noticed two posters conspicuously displayed in the window. One titled "Zero Tolerance for Canadian Wolves" highlighted a vicious-looking wolf baring its teeth and made claims that wolves were an invasive species that had decimated elk populations in neighboring Idaho and were about to do the same here in Oregon. The second, for the "Wallowa County Wolf Defense Fund," featured a dead sheep and described a fund set up to provide money for legal defense of people who shoot wolves in "personal defense" or in "defense of personal property" in the county. When I checked out with my two maps, I noticed bumper stickers for sale sitting by register which read, "Canadian wolves: smoke a pack a day" and "Canadian wolves: state-sponsored terrorism." I thought about asking the man behind the counter if he was concerned that the posters and bumper stickers might offend some of his potential clientele given the large number of outdoor recreationists who make their way through town, the majority of whom likely have positive associations with wolves. According to an ODFW poll of Oregon citizens, about 70 percent of all Oregonians, outdoors enthusiasts or otherwise, support the recovery of wolves. I would make a poor journalist, I realized, as the grim look on his face and a desire to avoid a scene deterred me, and I took my maps and went on my way.

Oregon's Imnaha pack occupies a territory that includes mountains in the Wallowas as well as the rolling grasslands and agricultural areas of the Zumwalt Prairie.

Since my first visit, my friend Al had made a number of forays to the area to hunt elk. A couple of years previously he had actually killed a bull, something very few people who hunt elk in Oregon manage to achieve. After his last trip he called me, and along with the stories of near misses and long walks, he told me about wolves—the Imnaha pack. He and his hunting mates had heard them howling, had seen tracks, and finally saw the wolves themselves. I could hear the excitement in his voice—all of a sudden a land that already seemed so wild just became a little wilder.

I was out to follow up on leads he had given me from what he had seen and what he had heard from other hunters. I was going in August, hoping to avoid the influx of people into the area for the hunting season. But rather than signs of wolves, everywhere I went I found cattle: cattle on the ridges, cattle in the forest, cattle in the streams, cattle on the roads. After several days of hiking and driving, I began to wonder how the wolves find deer and elk at all, having to sort their way through the throngs of cattle. Picking my way around cow pies to look for tracks in the mud by a small seeping spring, I wondered why cows aren't all that wolves eat, given the abundance of cattle on the landscape. The only tracks in and around the spring were of cattle. I didn't find a stitch of wolf sign on the entire trip.

What I found lacking in wolf sign that trip was more than made up for in the myriad opinions and stories I heard while out there and in the articles and websites I researched when I got home. The Imnaha pack's territory has put them in the center of the clash of evolving cultural values and divergent opinions about the management of our region's wildlands. This clash manifests itself in sensationalized stories about wolves, their behavior, and their impact on game animals, livestock, pets, people, and ecosystems.

From the letters to the editor in the *Wallowa Chieftain*, the local paper from the vicinity of the Imnaha pack, one might think that the wolves in Wallowa county were annihilating wild game and livestock. One Joseph, Oregon, resident wrote, "At the rate the deer and elk herds are dwindling here because of unchecked wolf, cougar and coyote populations, it won't be long until we have little or no hunting opportunities left in this area." Others claim that the wolves are part of a federal government plot to further deteriorate the rights of rural residents. According to a resident of Imnaha, Oregon, "The wolf was introduced in the West to drive the livestock producers out of business. Animal activists, allied with the government, are actively working to destroy the livestock industry. The Canadian gray wolf was introduced in the West illegally and with stolen federal funds under the guise of the Endangered Species Act. The wolf is another pawn in the game to drive rural resource users off of the land."

Conversely, if you read material put out by Defenders of Wildlife concerning the ecological value of wolves and related conservation issues, you wonder how a wolf can simultaneously initiate vast cascades of changes on a landscape through its predation on elk and at the same time have no significant impact on elk populations. In one publication Defenders of Wildlife claims that "By preventing large herbivores such as deer and elk from becoming overpopulated, wolves help maintain native biodiversity." Then, on the very next page, they state "Except when winters are extremely severe, the number of elk killed by wolves is not high enough to have a widespread biological impact on elk populations." Both statements cite peer-reviewed scientific research (something that most material produced by anti-wolf interests neglects to do) carried out in specific locations. However, obviously both these claims cannot possibly be true in all instances. Defenders of Wildlife has been among the most active national conservation groups advocating for the return of wolves to landscapes they have been extirpated from in the western United States. This organization has spearheaded initiatives designed to lessen the negative impacts of wolves on ranchers, including paying ranchers for livestock killed by wolves and providing training and materials for ranchers in methods that reduce the risks of wolf predation on livestock. Defenders of Wildlife has also been deeply engaged in the ongoing social debate about the role and value of wolves, presenting the animal in their publications in a light that supports the organization's efforts to restore and conserve the species.

WOLVES AND ELK

The Blue Mountains of Oregon and Washington are home to a large elk herd. In ecologically similar parts of the Rockies to the east, elk are the primary prey species for wolves. After many decades of increasing numbers, elk in the Blue Mountains have been declining in recent years. This decline has been linked to habitat succession and increasing predation pressure on calves from black bears and cougars, although human hunting has consistently been the largest overall source of mortality for adult elk in the area. Because of the declining numbers of elk, the ODFW has become more stringent about issuing hunting tags for elk here, much to the consternation of some of the state's elk hunters.

Historically elk appear to have been abundant in the Blue Mountains, but in the second half of the 1800s, intense human hunting dropped the population precipitously. During this time most elk hunting was market hunting, in which hunters would kill large numbers of elk to sell meat to the region's burgeoning cities. By the 1880s elk had become scarce not only in the Blue Mountains but across most of Oregon. In 1910 the state placed a total moratorium on hunting elk. Relieved of human hunting pressure and of most of their natural predators (wolves were close to extinct at this point, and mountain lions and bears were scarce) and supplemented by translocations

Two cow elk traveling at first light, heads up in an alert position.

of elk from elsewhere, elk populations rebounded over the next several decades. In the Blue Mountains ranching interests opposed reintroductions, and by 1924 the state was receiving complaints about elk competing with livestock. In 1933 the state reinitiated a more carefully regulated elk hunting season.

While the effects of changing plant communities, carnivore populations, and human hunting have been influencing elk populations in the Blue Mountains, it appears that the elk themselves have been shaping the landscape in return. During the summer of 2011, I joined botanist Mark Darrach to search for a four-acre ungulate exclosure— created by a stout, eight-foot-high fence designed to keep deer and elk out of the study plot that had been established decades earlier in the heart of the Blue Mountains—to measure these impacts.

The Sinks is a large tract of forests on a high-elevation mountain plateau in the Blue Mountains, including one of eastern Oregon's largest patches of old growth trees. With around sixty inches of precipitation annually, it is also among the wettest areas in the Blue Mountains. The elevation here is about 4500 feet, so much of this precipitation falls as snow in the winter. Though wolves have not yet been detected in the Sinks specifically, ODFW snow-tracking surveys regularly find tracks of the Wenaha pack, Oregon's second confirmed pack in recent history, on roads just a few miles away.

In the mid 1960s, hoping to understand the impact of elk browsing on forests in the Blue Mountains, researchers erected several fenced exclosures to keep hoofed mammals out of experimental plots of forest. Researchers monitored these plots for the next three decades, measuring changes in plant community structure inside and outside of the exclosures. Darrach had visited one of these exclosures around 2000 while inventorying the flora of the Sinks and recalled the stark difference between the vegetation inside and outside of the fencing. The official data collection and research at this plot had ended some years before our visit, and we could find no one who knew the current condition of the exclosure. We set out to find it with some questionable GPS coordinates and Darrach's vague recollections of its location.

Besides being a botanist specializing in rare plants, Darrach is also a geologist by training and an exceptional naturalist in general. As he navigated the maze of logging roads and spurs, he recounted stories from past fieldwork in the Sinks, an area of strange and perhaps supernatural happenings according to some. Darrach recounted that the Sinks is of significant cultural importance to the Confederated Tribes of the Umatilla who, in the early 1990s, successfully petitioned the U.S. Forest Service to cease logging operations in the area. The geologist in Darrach shone through as he explained how massive ancient landslides occurred here millennia ago, giving the area the rumpled topography it has today and hence the name, the Sinks.

We followed GPS coordinates as far as we were able, leaving the primary road for an unmaintained spur and eventually pulling off at a wide spot and parking. Stepping off the logging road, Darrach pointed out signs of intensive elk browsing on the forest landscape. In the sunny openings throughout the forest, a sea of waist-high western coneflowers dominated the vegetation. Darrach noted that coneflowers are not palatable to deer, elk, or cattle, and in places with heavy browsing pressure they tend to increase in abundance, as their competition is mowed down by ungulates but they are left untouched. Besides this subtle clue, several western yew saplings and a lone Rocky Mountain maple showed the distinctive stunted structure of heavy browsing pressure.

The lack of understory shrub cover in forests such as this in the Blue Mountains has been linked to intensive browsing pressure by elk, deer, and livestock.

Under the canopy of large grand and Douglas firs, the forest floor opened up. We wandered well-worn elk trails, noting the distinct pattern of dense coneflower patches in openings and an open forest floor, primarily devoid of shrub cover, under the forest canopy. Everywhere we went the tracks and other signs of elk abounded. After close to an hour of searching, we eventually stumbled into the exclosure. What we found was indeed striking.

Darrach pointed out several species of shrubs that were absent from the surrounding forest and, in his experience as a botanist in the Blue Mountains, generally uncommon and when found, often stunted from repetitive browsing. Several black cottonwood trees towered into the forest canopy, Scoular's willows grew over twenty feet tall in the understory, and thimbleberry shrubs grew thickly along the sunny edges where the fencing had kept hoofed mammals at bay for decades.

Because of the heavy snowfall during the winter here, every fall during the study period most of the fencing was taken down after deer and elk had migrated to lower elevations, then erected again in the spring as soon as the snowpack allowed. Since the official research ended, the fence has been left down, and the deer and elk have had access to the interior of the plot. With the exception of the cottonwood trees that had grown beyond where elk could reach their branches, many of the plants showed heavy browsing pressure from the past several years, but the straight stems and tall stature of these shrubs belied the freedom from browsing pressure they had grown up in.

In fact, a rigorous three-decade-long study of this and several other exclosures documented the effects of foraging by elk, deer, and livestock on Blue Mountains forests. At the exclosure I visited with Darrach, there was no record of any use of the area by either sheep or cattle during the entire duration of the study. Elk and deer used the area during an approximately six-month-long snow-free period each year. As we had noted anecdotally, the study concluded that ungulate browsing decreased the overall shrub cover and the diversity of shrub species outside of the exclosure. Rocky Mountain maple, Pacific yew, mountain ash, and serviceberry were among the shrubs most affected. A more subtle discovery was that herbivory decreased nitrogen accretion following disturbances (logging and fire) as ungulates selectively feed on key pioneer nitrogen-fixing plant species such as ceanothus.

Elk and deer, like humans, apparently have a propensity for eating the icing first. Through their selective foraging on preferred plants, hoofed mammals in great enough numbers can change the structure of forests. How much of this effect is related to the decline or absence of some of their traditional predators in this area, including wolves? And what happens to the elk when wolves return to a landscape like this one in the Blues? Heading east into central Idaho, where wolves were reintroduced in 1995, sheds some light on these questions.

Idaho's Bear Valley wolf pack roams a landscape of rolling mountains, dense lodgepole pine forests, and expansive mountain prairies and wet meadows in the Salmon River Mountains of central Idaho. Much of their home range is included in the Frank Church–River of No Return Wilderness, the largest designated wilderness area in the continental United States. It is a land of stark contrasts and fascinating sounds. In summer, sandhill cranes arrive from the south to rear their young in the middle of the vast wet meadows, their eerie pterodactyl-like calls echoing through the evening air. Cold, clear, star-filled nights often bring a blanket of morning mist over valley bottoms and slow-moving streams that snake through the mountains. Heading out at first light is the best way to find and observe wildlife in the open meadows. The high-pitched calls of cow elk communicating with their calves echo through the mist.

On one such morning I was out with a group of students. Eventually the herd of elk we were observing became aware of us. In the dim morning twilight, through the mist, we watched as a group of well over a hundred elk switched from an uncoordinated bunch of animals aimlessly grazing in the middle of a mile-wide meadow into a tight-knit unit which thundered away into the forest at the meadow's far edge. The ground shook beneath our feet, and when the elk disappeared from sight, their excited calls drifted faintly from the forest.

Cool summer mornings give way to searing sunshine and then booming evening thunderstorms which roll across the mountains, occasionally igniting wildfires in the dense stands of lodgepole pines which dominate the landscape. And of course, there are the wolves. The Bear Valley pack was the first pack I ever heard chorus howl, the long low call of an adult followed by other adults and then a cacophony of pups erupting into a jubilant family song.

The Bear Valley pack's home range is close to where the USFWS released wolves in 1995. From this central location wolves dispersed across the mountains far and wide. During the summer of 2009, I took several students in search of the Bear Valley pack's rendezvous site. We established a small camp in the Frank Church Wilderness in a location where we had detected repeated activity the week before. We awoke in the middle of our first night to wolves howling in the distance. In the morning we plotted onto a map where we thought the howling was coming from. Between the seven of us we isolated two basic directions. One pointed into a meadow system about four miles to the northwest, the other to another meadow system a similar distance due north. We decided to head to the northwest that day and see what we could find. On the trail into the meadow we discovered older sign of wolves, tracks baked into now dry mud and old scats. The meadow itself was beautiful and contained more older sign of wolves and elk but nothing fresh.

The next day we set out to the north. Shortly out of camp we passed the remains of a deer, killed and consumed by wolves, which several of my students had discovered the previous week. We paused to see what progress had been made on it. The students must have discovered it very shortly after it had been killed as the carcass had been mainly intact. Now, many days later, the carcass was reduced to scattered bones, tufts of hair, and a large pile of partially digested vegetation, the stomach contents of the deer. Scattered around the area were little piles of crunched bone fragments.

Carrying on northward along a game trail which faded in and out, we picked up wolf tracks also heading north in dusty patches of the trail. The tracks of several animals were fresh from that night. Farther on, we found a fresh wolf scat, loose, tarry, and stinky. Such scats general indicate the wolf had been feeding on a fresh kill. The first feeding focuses on the internal organs and largest muscle groups. Little hair, skin, or bone is consumed, and the resulting scat is runny and loose. Clearly the pack had made another kill since the deer we had inspected.

We stalked down into the meadow which we had located on the map earlier, a vast circular grassland stretching more than four miles across and bisected by several islands of trees. Through binoculars we made out a large herd of elk on the far eastern edge of the meadow and several pairs of sandhill cranes. According to the map, a pair of streams oxbowed through the meadow with several marshy sections, a perfect spot for a rendezvous location, with ample food and water, tucked away far from any traveled human trails. We spread out and moved slowly across the meadow to a stretch of trees and from there worked our way east. More fresh tracks and scats indicated that indeed we were close. At the end of the small patch of trees, several of my students caught a glimpse of a large black wolf which rose and moved along the distant treeline. We collected at the edge of the trees, sat, and watched the elk and cranes, hoping to catch another glimpse of a member of the pack, but all was quiet. We retreated, content that we had located the general area of their current rendezvous site.

Several years later, when I returned to the area I encountered the pack very close to where I had camped years earlier and had the opportunity to watch and listen to members of the pack for several days. The last sighting I had of them was at night. Howling awoke me from my sleep and I peered out from my tent in time to catch the silhouette of a wolf moving in the moonlight out in meadow next to the patch of trees where my camp was.

LEFT: A cow and calf elk graze in a misty meadow in the Salmon River Mountains of central Idaho.

OPPOSITE BELOW: A wolf pup from Idaho's Bear Valley pack pauses on a fallen log before disappearing into the forest.

BELOW: An adult from the Bear Valley pack howls from across a wet meadow, close to the pack's rendezvous site at the time. Salmon River Mountains of central Idaho.

The return of wolves to the mountains in central Idaho evoked a great deal of concern about the potentially devastating impacts wolves would have on the state's elk population. Wolves like to eat elk and so do people—the stage was set for a classic case of competition between two large carnivores. Further complicating the issue is that people not only like to eat elk—they make a lot of money from them, too. Elk is probably the state's most prized game species, and hunting in general is an important part of the central Idaho economy. Furthermore, about 46 percent of the Idaho Department of Fish and Game's (IDFG) entire budget (its largest single funding source, and nearly all of its discretionary funds) comes from hunting license and tag sales. Clearly for many in Idaho, there is a vested interest in maintaining the state's elk numbers and human hunting opportunities.

Tension between conservation and consumption has a long history in North America's hunting community. Following the decimation of many game species (often by market hunting), sport hunting interests drove many conservation efforts for game species, including elk, in the early 1900s in North America. Conservation-minded hunters are among the most important roots of the continent's modern conservation and environmental movement, and many contemporary hunters continue to be staunch conservationists. These efforts led to the protection of millions of acres of habitat, including the creation of numerous national parks and modern game-management techniques that allowed for the resurgence of game species populations across the continent in the twentieth century. The present abundance of elk, which created excellent conditions for wolves in much of Idaho, can be traced back, in large part, to these hunter-driven conservation efforts. Hunter-based organizations such as the Rocky Mountain Elk Foundation and Ducks Unlimited continue to move forward habitat conservation and play an invaluable role in contemporary conservation efforts. Conversely, efforts to increase game species have a history of being narrowly focused on the specific desirable species and have often included predator control and other tactics that have detrimental impacts to overall ecological integrity.

The IDFG sets specific goals for elk populations across the state and then carefully tracks them in order to set hunting regulations accordingly. According to the IDFG, humans account for most of the annual mortality of elk greater than one year old in the state. The IDFG also reports that now, almost twenty years after wolves have returned to Idaho, most of the state is either at or above its targets for elk populations. However, the IDFG has also identified a few areas where elk populations are below their targets or are in significant decline. Long-term changes in habitat are seen as a primary driving factor. Large portions of north-central Idaho burned in intense stand-replacing fires in the early 1900s. Now, with natural forest succession, these lands have largely returned to forest from open shrublands. Predators also influence elk populations in Idaho. A broad study of predation on elk calves across the Northwest found that the combined influences of multiple predators, including mountain lions, black bears, grizzly bears, and wolves, can lead to declines in elk populations, with bears found to be the most significant single factor. In locations with more than two of these four predator species, declines were more likely. In a few specific locations within the state, the IDFG has linked a steep decline of elk numbers to predation primarily from wolves.

The Lolo elk management zone experienced a drop in elk numbers from more than 5000 in 2006 to fewer than 2500 in 2010, a greater than 50 percent decrease. The state considers wolf predation to be the primary source of mortality in this herd, though this decline also appears to be the tail end of a population decrease that started with conversion of habitat from open shrublands to forests. At its peak in the 1980s, this herd numbered sixteen thousand animals. The IDFG, in an attempt to increase elk numbers in this area, set targets of removing most of the wolves from this zone.

The finding that wolves were the primary source of mortality for elk in the Lolo zone became a rallying point for those interested in reducing wolf numbers to "save the elk" (so that humans can hunt them). In 2010 the IDFG distributed a news release written by their director, Cal Groen, which was quoted extensively by media far and wide covering this topic. He stated: "The elk situation in the Lolo elk management zone didn't happen overnight. The Lolo elk herd had glory days after major fires in the early 1900s created phenomenal elk habitat. . . . But regrowth of brush and forest turned great elk habitat into poor habitat. Predation by bears and mountain lions took its toll. Following the severe winter of 1996–1997, Lolo elk numbers dropped by nearly half." After a brief respite in the ongoing drop in elk numbers, "wolves took over and became the leading cause of Lolo elk deaths. It wasn't until May of last year that the state could finally manage wolves" (because of their status as federally endangered). And here we get the line that was picked up and run with by the press: "By then, the balance of elk and wolves in the Lolo Zone was completely out of whack." In this statement, the IDFG departs from a scientific assessment of fact and enters into a murky world of opinions and competing human interests. What is clear from these statements is that the IDFG would like more elk in this area and that they see wolves as an obstacle to this goal.

The news release ends with another impassioned declaration: "Even with few wolves, changes in the landscape make it unlikely Lolo elk will return to the all-time highs of the 1980s. But Fish and Game will do what it takes to restore the health of the Lolo herd. For many of us, it's more than just professional interest; this herd has personal significance to many Idaho wildlife managers."

What exactly does "health" refer to? Habitat succession is a natural phenomenon, as is predation. Beginning with habitat succession and furthered by native carnivores, the elk population here has undergone a predictable drop in numbers. What is unhealthy about this? Without any human interventions in this area, the elk population would likely stay depressed until predation decreased (with decreases in predator numbers because of low prey abundance) and favorable habitat changes once again increased its carrying capacity. That wolves could contribute to a continued decline in human hunting opportunities in the area is very likely. The notion that wolves would completely eradicate all the elk in the area and then prevent them from ever returning irrespective of habitat changes over time is baseless from a scientific perspective. That humans, a hunter of large ungulates, would feel threatened by the presence of a competing carnivore on the landscape, and therefore would attempt to reduce this sense of threat and competition, is completely predictable. But being the complicated and confusing species that we are, humans have taken a basic issue of competition between species with a similar niche and complicated it by adding a quasi-moralistic stance involving an imperative to maintain "balance" and improve "health." This is especially interesting since, as we will see, other humans, with an opposing agenda, use the same terms to justify maintaining or restoring wolf populations. We are an odd species.

The management unit that includes the Bear Valley pack is another area below the state's elk population targets and where the IDFG has cited wolves as the primary

Lightning strikes started this fire burning in the home range of Idaho's Bear Valley pack. An earlier fire burned the landscape in the foreground, producing better habitat for elk, much to the benefit of wolves and human hunters alike.

source of elk mortality, though the influence of bear and cougars on the elk herd is "mostly unknown" according to a 2010 progress report. Parts of this management zone suffer from poor range quality because of past overgrazing by ungulates, while large wildfires in other areas have increased the habitat value for elk in recent decades.

Change is definitely afoot in the landscape which the Bear Valley pack calls home. For over a century much of this land was heavily grazed each summer by sheep and cattle. In the 1990s all the sheep allotments—public land leased by the federal government to ranchers for livestock grazing—were closed following a decline in the sheep industry which made their use uneconomical. The cattle allotments were closed between 1999 and 2001 as part of attempts to restore salmon runs in the area. In just the brief seven years that I have been visiting this area, I've watched the landscape slowly changing. The drier parts of the meadows are still in recovery with grasses and hardy pioneering plants slowly recolonizing patches of open dirt. Along stream courses that had been denuded by grazing sheep and elk, willows have returned and thickened by the year. Beavers appear to be increasing in distribution. The lurking forms of chinook salmon haunt the clear waters of the streams. Having traveled hundreds of miles from the sea to spawn, the large fish are another beneficiary of the resurgent riparian communities here.

The connection between wolves (and other apex carnivores), elk (and other wild hoofed prey), and riparian zones (and other forest communities) has been the focus of a great deal of research and pontification. In the Canadian and United States Rockies, the return of wolves to landscapes from which they had been absent has provided the opportunity to see specifically how the absence and return of this carnivore affects ungulates and in turn many other parts of an ecosystem. One study found that the extirpation of wolves and grizzly bears led to increases in moose populations and subsequently a decline in riparian plant communities and songbird diversity. Another study found that in areas with wolves, changes in elk use of the landscape allowed for increases in riparian vegetation that in turn allowed beaver populations to increase. Most famously, in Yellowstone researchers documented resurgence in riparian zones, cottonwoods, and aspen forests and correlated this to the return of wolves, though the exact mix of contributing ecological influences is still being debated.

Those who champion wolves and other large carnivores as important for the conservation and restoration of natural ecosystems look to these findings as evidence that wolves are vital for "maintaining the health of ecosystems," as Defenders of Wildlife has put it. This sounds like a familiar stance. Here the term *health* refers to a benefit from wolves killing elk, while wolves killing elk has also been characterized as a detriment to health. Whose health are we talking about here really—an ecosystem's, the elk's, wolves', or that of humans?

If we see wolves as a species competing with us for food or threatening our livelihood, it makes sense that we would perceive them to be a threat to our "health."

When we start looking at wolves in a broader context—as in how they fit into a larger ecological community—we can again find some basic underlying human needs that people may be striving to preserve.

As modern western thinking has moved toward a more holistic understanding of how natural systems function, there is an increasing appreciation that things like biodiversity and landscapes that conserve and circulate water and nutrients effectively underlie the continued ability of ecosystems to provide natural services, such as clean air and clean water, upon which humans all depend. While these connections to human health may not seem as directly linked to our survival and security in the moment as dealing with the threat of a competing carnivore, in the long run they may be what is vital to our survival as a species in the modern world. Mix this in with a bit of shame over what we as a species have done to the region's natural heritage, and once again we discover a basic biological urge for survival now intermixed with a sense of moral imperative to protect and restore what we have defiled. That we are conscious of our impacts and influences on ecosystems creates a uniquely human quandary. Wolves kill and eat elk. Humans kill and eat elk, and also pontificate on whether or how or how many they should kill, as well as who else should get to kill them.

Emily Gibson inspects the remains of an elk killed by wolves she discovered among the willows on the edge of a shallow stream in the Salmon River Mountains.

Of course, not everyone who hunts elk is disturbed by the presence of wolves on the landscape. Many modern hunters see hunting as both an endeavor to procure food for the family and an opportunity to connect with the wild and carry on a cultural tradition that dates back untold generations. My friend Al Thieme, whose scouting for hunting season first drew me to the Wallowas, reflected on this from his own experiences hunting elk there: "As a hunter, the existence of wolves in the area we were hunting in enlivened the experience for me. . . . How amazing for our hunting party to be able to interact with these wild creatures in their own home range. It is lamentable that most people in the modern world have such little exposure to large wild mammals."

WOLVES AND LIVESTOCK

Understanding the relationship between people and livestock and the impacts of livestock on Pacific Northwest ecosystems is not simple, but it underlies much of the contemporary social tension about the place of wolves in the region.

Cattle, such as these sharing the landscape with the Imnaha pack, are ubiquitous across much of the public lands of northeastern Oregon during the summer months. One study found that cattle seek out roads in response to the immediate presence of wolves in an area, possibly using human presence as a shield from the predators.

Similarly, the history of livestock production in the Pacific Northwest has had a number of ecological impacts on wildlands that directly or indirectly relate to wolves. Amazingly, European livestock—horses—arrived in parts of the Pacific

Northwest before European explorers and settlers did. By the early 1700s the Nez Perce, whose territory included present-day eastern Washington, Oregon, and western Idaho, had amassed huge herds of horses, originating from animals they had traded for or had stolen from tribes to the south. Originally brought to the American southwest by the Spanish, horses then spread through established tribal trade networks, quickly becoming an integral part of Columbia Plateau tribal life and a symbol of wealth. They also undoubtedly initiated significant environmental changes in the landscape by their presence, as well as through the shifts that access to horses created in the material lives of the plateau tribes.

European-American settlements introduced cattle and sheep in the region from the coast to the interior. By the middle of the nineteenth century, livestock grazing was well-established here as homesteaders and ranchers drove cattle into just about every low-elevation meadow and forest opening even in the most remote areas, while large herds of sheep and cattle were moved up to high-elevation meadows in all the mountains of the region. In 1867 the first wagon road over Snoqualmie Pass, the lowest pass in the Washington Cascades (through which Interstate 90 now runs), was constructed and used primarily to drive cattle from the Ellensburg area to Puget Sound to market.

Along with homesteading, large-scale commercial livestock operations began in the interior in the mid to late 1800s. In both instances, but especially with large-scale business operations, predator control was seen as vital for economic viability. This period of regional livestock increases accompanied or followed the decimation of native ungulate herds by hunting, both market hunting for meat to sell to growing population centers and subsistence hunting by settlers.

To increase their efficiency, livestock producers, from small-scale homesteads to huge livestock barons, attempted to dismantle or counteract natural forces that otherwise limit populations of hoofed mammals, including killing native predators and supplementing winter feed to maintain abnormally high population levels. Today Washington State estimates it has a population of about 1.1 to 1.2 million cattle and 46,000 to 58,000 sheep. Oregon estimates a somewhat larger livestock population at about 1.4 million cattle and 217,000 sheep. Regionally, livestock are grazed on both private and public lands. In Washington 3.36 million acres of public lands are grazed by private livestock. In both Washington and Oregon most public-lands grazing occurs in the eastern portions of the state. Annually about two-thirds of all beef cattle in eastern Oregon are grazed on public lands.

Grazing on public lands administered by the USFS and the Bureau of Land Management is regulated by a permit system developed in the 1930s, in part to deal with deteriorating conditions caused by overgrazing. However, because of the political influence of ranching interests, rather than protecting fragile rangelands from abuse, the Taylor Grazing Act of 1934 has institutionalized a system that provides tremendous subsidies to ranchers in the form of below-market-rate grazing permits and has made it even more difficult to reduce the impacts of overstocking of public lands. The value of these subsidies is so great that ranchers can get bank loans against the value of their grazing permits based on the difference between the permits' market value and the minimal price charged by the federal government for them. In many areas of the Blue Mountains and other interior landscapes in the region, the cumulative impacts of chronic overgrazing and predator control are readily apparent.

The North Fork of the Burnt River drains a section of arid mountains south and east of the Imnaha pack's territory. Dominated by arid ponderosa pine and Douglas fir forests, it is a familiar and typical landscape for this part of the Pacific Northwest. The river itself is still in the relatively early stages of recovery from nineteenth-century placer mining operations for gold that literally turned the river bottom inside out, destroying the riparian corridor in the process. The river's watershed ownership reflects a pattern common across the West, an artifact of the homesteading era and historic mining claims. Along the choicest sections of the river are relatively small inholdings of private land surrounded by vast stretches of public lands, in this case national forest.

In the Pacific Northwest interior, river corridors and streamside forests are usually the most biologically rich and diverse part of the landscape. But wandering along the banks of the Burnt River, I found a very different story. On a hot summer day in August with the smell of cattle wafting in the air, it was readily apparent that the river was also the most scarred part of this landscape.

I picked my way alternately between sections of river deformed from mining operations and deeply eroded oxbowed sections nearly devoid of riparian vegetation. Streams receive disproportionately high use by cattle, attracted to the water and riparian vegetation for forage. Where willows, cottonwoods, and an entire suite of understory shrubs and forbs should line the banks of the river, here only the occasional stunted alder clump has managed to persist. Alder is not a palatable browse plant for ungulates so it survives after the tastier willow and cottonwood have disappeared. Once everything else is gone, even the alder is consumed: here the alders are dwarfed, sculpted into the distinct bonsai-like appearance created by heavy browsing pressure over many years. In some places the eroded stream banks were eight feet deep and devoid of trees and shrubs. Instead of the quick darting of trout in the shaded cool waters of a stream with an intact riparian zone, algae bloomed in the eddies along the edge of the slow-moving trickle of warm water. Cow feces in and along the stream suggested the source for the excessive nitrogen in the water that leads to such algae blooms. The complete lack of shade and low water levels, caused by erosion which drops the entire system's water table, have completely destroyed the stream as trout or salmon habitat. The lack of riparian vegetation creates a major obstacle to beavers' recolonization of this particular reach of the river as they would have nothing to eat. This is a double ecological

insult as beavers are an excellent natural balm to ailing streams, slowing erosive floodwaters, increasing the width of the riparian corridor, and raising the water table.

Much of the Burnt River's watershed is included in active grazing allotments that are the focus of a lawsuit brought by conservation groups, including Hells Canyon Preservation Council (which has also been deeply involved with wolf conservation efforts in the area), against the USFS.

Two cows graze on the banks of the North Fork of the Burnt River in northeastern Oregon, here deeply eroded and almost completely devoid of riparian vegetation, both signs of chronic overgrazing.

In 2005 the U.S. Congress passed legislation which allowed the USFS, under certain circumstances, to bypass all the regulations of the National Environmental Policy Act in renewing grazing permits. Without this legislation, such renewals are subject to lengthy reviews of the environmental impacts of continued grazing on the public lands allotment. To qualify for exclusion from this review process, the U.S. Forest Service had to demonstrate that the ongoing grazing was meeting or moving toward the agency's ecological objectives and that there would be no significant impacts to imperiled species or sensitive habitats.

In this case, the plaintiffs claim that many of the grazing allotments in northeastern Oregon that were renewed without any environmental assessment do contain such sensitive species or habitats, according to the USFS's own assessment. Further, they claim that there was no monitoring data to support the premise that grazing was not impeding the agency's ecological goals. A cursory inspection of the North Fork of the Burnt River and other grazing allotments in question lends credence to these claims. The Forest Service's apparent attempts to evade its own assessments and undermine its own environmental targets in order to move through grazing allotment renewal more quickly has been seen by conservation interests as an example of the political influence of the livestock industry, an idea further supported by the fact that the U.S. Congress got involved with this at all. Those who favor continued grazing on public lands often characterize such moves as attempts to support local economies and deal with inefficient federal bureaucracy.

The impacts of overgrazing are not limited to the banks of the Burnt River in Oregon. They are ubiquitous across public wildlands throughout the region's interior. Aspen stands are another place to find more signs of this. Unlike in the southern Rockies of Colorado, New Mexico, or Arizona, aspen groves in the Pacific Northwest are never a dominant forest type—instead, small groves nestle into moist ravines or dot hillsides around a seep or spring. These rare patches of broadleaf trees in a sea of conifer forests are havens of increased biological diversity, perhaps most readily apparent in the spring when the songs, breeding plumage, and nesting activity of a wide array of birds decorate the silver-green stands of trees. That these groves attract special attention from cattle is also often readily apparent.

West and north of the Wallowa Mountains, across the Columbia River basin, jutting east out of the Cascade Range in central Washington, a long chain of rolling foothills called Clockum Ridge descends to the Columbia River. On a matrix of state, federal, and private land, forests of ponderosa pine and Douglas fir stretch out across the rolling hills mixed with open shrub-steppe. In the fall golden larch needles speckle the steep-sided drainages, and deer and elk descend from the Cascades to winter in the area, drawn by the smaller snowpack, milder weather, and abundant bitterbrush on south-facing slopes. Spring comes earlier here than in the higher mountains and shortly after the flush of balsamroot flowers come the cattle, brought to their summer range by ranchers with grazing permits for these public lands. The cattle linger until the fall when they are replaced by hunters searching for deer and elk.

On a crisp, windy fall day exploring this area, I came across a small patch of aspen tucked into a slight depression on a broad ridgeline. Rushes and other water-loving plants hinted at the higher water table here. Always attracted to aspen groves myself, I gravitated into the stand to look for bird nest cavities and scars from black bears climbing the smooth-barked trees. As I entered the grove I noted a recruitment gap—a lack of young trees—which is typical in many of the region's aspen stands. The stand comprised large mature trees, standing and fallen dead trees, and stunted saplings with the scars of repetitive browsing. Missing were any young trees larger than the stunted saplings. For many decades not a single sapling had been able to grow above the height at which deer, elk, and cattle could reach.

Without something changing, groves of aspens such as this one basically have a death sentence. Aspen spread through their roots, and a small stand of aspens is likely a single plant referred to as a clone. Each trunk is a single stem of one organism that is connected underground through the intertangled root system. If none of the younger stems are able to reach maturity, eventually the older stems die and the aspen grove disappears.

In this particular aspen stand, the telltale pattern of intensive use by cattle was also apparent: cow-pies, tracks in the mud, mashed vegetation where the beasts lazed away the hottest parts of summer days in the cool shade of the aspens. Mixed in with this pattern were the tracks and scats of elk. Who was responsible for the condition of this aspen stand and what will its fate be? In some parts of the Rockies, the return of wolves coincided with a resurgence in aspen regeneration. But this response has been far from uniform. In areas with heavy livestock use in the Rockies, wolves are often killed to reduce losses to ranchers. Such removals, by design, preclude wolves from influencing the impacts of livestock.

While the ecological impacts of elk herds unleashed from predation pressure have subtly altered various forest communities, the impact of livestock grazing across much of the interior portion of the region might be better described as a biological holocaust. Livestock production is the most widespread commercial use of arid wildlands in the Pacific Northwest. Decades of research on livestock production on public wildlands has created an overwhelming collection of data showing the wide-ranging detrimental effects of livestock on the ecological integrity of ecosystems in our region. Stream-bank erosion, soil compaction, expansion of invasive species, increased stream pollution, increases in water temperature in streams, increased bacteria counts in surface water, lowered water tables, decreased vegetation cover, and the decline of sensitive bird, amphibian, and mammal species have all been linked to cattle grazing in the Northwest.

The smooth bark and shimmering leaves of aspen stands attract the attention of humans, while the soft wood attracts cavity-nesting birds, and a variety of mammals favor palatable leaves and buds. For decades, intensive browsing pressure from elk or livestock has prevented many aspen stands, such as this one in the Wallowa Mountains, from successfully recruiting saplings into mature trees.

Woodpeckers, such as this pileated woodpecker in northwestern Montana, seek out aspens to construct nest cavities which are often subsequently used by many other species.

Making the world safe for cows and sheep has long been a goal of western civilization. Wherever Europeans have gone, from tropical rainforests to boreal taiga, they have brought their domestic animals with them and worked hard to create an idyllic environment for them. To this end, livestock production is typically accompanied by significant alterations to landscapes, such as road construction, fencing, modifications of natural springs, and predator control. Many of these alternations tend to reduce the ecological integrity of the landscapes over time. The goals of habitat modifications and predator control are to reduce the limiting factors on a landscape's carrying capacity for livestock in order to maximize the economic gain of the endeavor. Stocking very high numbers of livestock on a consistent basis creates conditions similar to those caused by an eruptive deer or elk herd, while predator removal artificially interferes with the natural feedback loops which limit this state.

Predator control, the original cause of extirpation of wolves in the region, has continued to the present time in the United States, at the cost of thousands of native carnivores yearly at a direct financial cost to the public of millions of dollars annually. In just 2010, USDA Wildlife Services, the U.S. federal agency responsible for predator control activities associated with livestock depredations and human health and safety concerns, intentionally killed 452 wolves, 80,639 coyotes, 2 grizzly bears, 578 black bears, 360 mountain lions, 1374 bobcats, 12,319 raccoons, 8253 striped skunks, 1 wolverine, 5 fishers, 117 river otters, 3 kit foxes, 21 swift foxes, 1706 gray foxes, 2130 red foxes, 1 golden eagle, 12 bald eagles, and 748 red-tailed hawks. In Idaho during 2010, 80 wolves were killed as a part of predator control activities, estimated at about 8 percent of the state's entire wolf population.

Amazingly, despite millions of dollars spent and tens of thousands of native predators killed, the effectiveness of predator control as a means of helping the livestock industry is questionable. One researcher, who looked at the effect of predator control efforts on coyotes and the impact of these efforts on the sheep industry from 1920 to 1998, declared that her results "suggest that government-subsidized predator control has failed to prevent the decline in the sheep industry and alternative support mechanisms need to be developed if the goal is to increase sheep production and not simply to kill carnivores."

The livestock industry's fear of wolves combined with their political clout has had significant impacts on the management of wolves in the U.S. Rocky Mountains, leading to the destruction of entire packs in areas of conflict between wolves and cattle. Wolf control activities designed to reduce conflict between wolves and livestock have directly limited any population of wolves from establishing in areas prioritized for livestock production. This choice to control wolf populations and distribution represents a continuation of a policy designed to maintain artificial conditions for raising livestock at the cost of ecological integrity. With the return of wolves to Oregon and Washington, the predator's impact on both states' livestock industries received a great deal of attention in the state management plans, including both an assessment of the potential cost of wolf depredation and a wide variety of tools for the state to use to prevent it or deal with it when it does occur.

The short history of what might have become another breeding pair of wolves in Oregon illustrates the initiation of a policy for managing wolf and livestock issues similar to what has been going on in Idaho, Montana, and Wyoming. In the spring of 2009, reports of wolves began trickling in from the hills around the Keating Valley, southwest of the Eagle Cap Wilderness (the Imnaha pack's home range is on the opposite side of the wilderness area). In mid April, over the course of two nights, these two wolves attacked and killed twenty-four lambs that were penned for the night on a ranch in the valley, a fact confirmed by the ODFW and USFWS. In response to this, one of the wolves was live-trapped and radio-collared and extensive deterrents were constructed at the ranch. Several months later in late August, the ODFW confirmed that these same wolves killed three sheep and a goat in a single evening on a different ranch in the valley. This repeated depredation following a series of nonlethal deterrent measures was the threshold at which point "lethal control" was warranted under the state's management plan. Despite being a state-listed endangered species at the time, both wolves were trapped and killed by federal Wildlife Services agents to prevent further depredations.

The plight of the Keating wolves demonstrates how imperative dealing with livestock depredation issues is for wildlife managers such as the ODFW. Killing the animal you are trying to restore would appear counterintuitive at first glance. However, many people consider killing wolves that kill livestock to be vital for the success of wolf recovery efforts. Killing wolves who are harassing or killing livestock is promoted from both a biological perspective—removing animals with a predisposition to this sort of behavior—and a human social one—appease livestock interests and mitigate the impacts of wolves on this small portion of society which bears a direct economic burden from the presence of wolves.

Oregon's and Washington's wolf management plans clearly state that the plans do not intend to address the issues of livestock grazing on public wildlands nor the ecological impacts of livestock grazing as it relates to wolves. Instead, both management plans focus on methods—including killing wolves under certain circumstances, as well as managing wolves and reimbursing producers for livestock killed by wolves—for reducing conflict and accommodating livestock interests.

While avoiding the issue of livestock grazing on public lands in wolf management plans has been politically expedient, it comes with ecological costs, including the loss of much of the potential value of the natural services provided by wolves. Essentially these management plans are designed to allow for the persistence of wolf populations in each state but to limit or exclude them in places where livestock production is prioritized, including many public lands. Even nonlethal deterrents, such as increasing human presence around livestock, would not only decrease predation on livestock by wolves but may also deter wolves from using these areas at all, which is ironic, as some of these landscapes would greatly benefit from a shift in predator-prey dynamics to counteract the long-term impacts of heavy grazing by hoofed mammals.

Ranching in many parts of the West has evolved from a primarily economic activity to one of tradition and social identity. Many ranching operations across the interior of the Northwest depend on heavy public subsidies in the form of cheap grazing permits for public land, federal predator control programs, range improvements carried out by the federal government, and federal agricultural tax subsidies. Despite this, as cattle prices have dipped in recent decades and land values have increased, economics have pushed many ranchers out of the business altogether. Several conservation groups, seeing an opportunity here, have negotiated deals to purchase grazing rights from ranchers and then have the allotments permanently retired by the land management agency.

Most conservation interests support efforts to reimburse ranchers for their losses. Why? One answer is that both wolves and cows are less valuable than the land they live on. As is the case with the Burnt River in northeastern Oregon, across much of the interior of the Northwest, a patchwork of land ownership which dates back to the days of homesteading has left low-elevation valley-bottom lands in large parcels owned privately while the mountains around them are owned by the public. Low-elevation and valley-bottom lands are both critical ecologically and highly desirable living places for humans. Real estate developers see money in subdividing working ranches into many smaller lots, building "ranchettes," and selling to urbanites looking for their getaway where they can escape the daily grind and enjoy the wonders of nature. Ironically, from a conservation perspective this can be a disaster. The landscape fragmentation that comes from such subdivisions and the conversion of open ranchland to more heavily populated and roaded landscapes degrades what is often vital winter habitat for wildlife. Most livestock operations that run cattle on public lands during the summer would not be viable without access to these grazing allotments on

The tracks of black bear, mule deer and wolf share a muddy patch on a road in the Wallowa Mountains.

public lands. As land values in many parts of the interior Pacific Northwest have increased and the economics of livestock production have become more precarious, the impetus for ranch owners to sell their property for development increases. Those interested in wolf and wildlands conservation may be left to choose between supporting the livestock industry and keeping these ecologically valuable lands as open space, or dealing with the development of these lands and a greatly increased human footprint on the landscape to the even greater detriment of the ecological integrity of the region.

It wasn't until 2011 that I managed to find signs of wolves in the Blue Mountains. Following up on a tip from a local biologist, I was poking around again east of the Eagle Cap Wilderness, dodging cattle on the roads as I went. On my second morning out in the field, I picked up fresh wolf tracks in the dust of a well-used ATV trail. A lone animal had cut onto the trail the night before. It had walked down the trail a short way, scent marked on the side of it, and darted off again back into dense lodgepole pine forest, heading downstream in the valley that the ATV trail was traversing. I walked back to the road where my truck was parked, figuring I would drive down the valley in hope of finding where the wolf might have cut back onto the road.

The dirt road yielded hundreds of cattle tracks. Assuming that the wolves' tracks would have been obliterated by the cattle, I carried on down the road, eventually coming across the livestock ambling along. They scuttled off both sides, several running down the road ahead of my vehicle for a couple hundred yards before cutting off. Interestingly, a study from the Canadian Rockies on the response of cattle to the presence of wolves in the immediate vicinity noted that cattle did not immediately respond to the predator's presence but sought out roads and trails after wolves departed, possibly using the increased presence of humans as a shield. Researchers found a very different response from elk, which quickly sought out forest cover instead.

I drove down a little way and pulled off to look for tracks along the untrampled section of road. Just as I stepped out of the vehicle howls echoed through the valley, multiple wolves calling from what seemed to be farther downstream and up the valley's north side. I stopped and listened to the eerie sound carry over the pine trees, the cattle, and the stream. There were no tracks to be found on the road, and shortly the cattle came around the bend in the road, apparently unperturbed by the wolves' howling.

One might naturally assume that wolves are usually predatory toward livestock. Given their reduced survival instincts and fitness, domestic animals would appear to be perfect candidates for predation by wolves who are opportunistic by nature. However, in actuality, wolves demonstrate a wide variety of behaviors in the presence of livestock, from indifference to active predation. Packs, such as Oregon's

Imnaha pack, that inhabit areas with extensive livestock grazing during the spring denning and summer pup-rearing season and whose winter habitat overlaps with private lands where cattle and sheep are kept through the winter months, do kill livestock from time to time. However, livestock are rarely the primary food source of these wolves.

Wolf-caused mortality on cattle has not been found to be significant on a broad scale in the Rockies, and estimates for a recovered population in Oregon and Washington predict a similar situation in these two states. A study of the economic costs of wolf predation on the livestock industries of Idaho, Montana, and Wyoming through 2003 estimated that losses averaged $11,076 or about 0.01 percent of the gross income for the industry in the region, though the study noted a consistent increase in these costs as wolf populations expanded during the study period. Idaho's population of adult domestic sheep averages about 231,000. In 2010 in Idaho, with a population of 87 packs and a minimum of 705 wolves, Wildlife Services confirmed that 75 cattle and 148 sheep were killed by wolves. It is important to note that confirmed kills are an underestimate of what wolves actually kill as not all carcasses are recovered or conclusive as to the killer. Actual numbers might be two, three, or even six times higher than this. Tripling the confirmed number of sheep killed by wolves, wolves would still be responsible for the demise of less than one-fifth of 1 percent of the state's sheep population in 2010. For comparison, up to 8 percent of the entire state's wolf population was killed specifically in efforts to mitigate livestock depredations. In the northern Rockies, domestic dogs consistently kill many more sheep and cattle than wolves do.

It is unclear how much wolf control efforts have reduced losses to the livestock industry. Further, despite these efforts, in some localized areas losses for specific ranchers can be considerable. The amount of wolf predation on livestock is clearly linked to specific grazing methods, including the type of terrain livestock are put on, the level of supervision, and the type and age of livestock. Sheep are more likely to be targeted by wolves, and calves are more vulnerable than adult cattle. In Montana and Idaho, wolf predation on cattle was more likely in pastures farther from human residences, in areas with a larger number of elk sharing the range, and in areas with greater brush cover. Not surprisingly, cattle located closer to active wolf den or rendezvous sites are more at risk to predation than those pastured at a greater distance. Besides removal of depredating wolves, other efforts to minimize livestock losses have included modified grazing practices such as greater supervision of livestock through range riders and the use of guard dogs.

Sometimes wolves will kill more than one prey animal at once, and more than they can readily consume at once, a behavior commonly referred to as surplus-killing. While documented occasionally with wild prey, it is most commonly found with sheep and domestic livestock, perhaps in response to domesticated animals' lack of defensive strategies. With wild prey, surplus killing may be associated with unusual conditions that make killing considerably easier, such as snow conditions in which hoofed animals penetrate the snow deeply but wolves can travel on top of the snow's crust. Wolves may instinctively kill prey when it is relatively easy to do so.

When wolves surplus-kill domestic livestock, they are often prohibited from returning to feed on the prey in order to deter them from killing livestock again. However, animals have been documented to return to surplus kills of wild game when other food sources disappear, sometimes months later. Although likely motivated by instinct, surplus killing, especially of livestock, has often been used as evidence of the wasteful and evil ways of wolves, in contrast to our culture's espoused values of thrift, taking only what we need, and avoiding causing pain or taking life unnecessarily, a little ironic coming from a species with a track record such as ours in these matters.

The apparently paradoxical biological and social drives of various segments of our culture have created a polarized and contentious atmosphere when it comes to wolves and wildlands. Are the biological drives to survive and bend our surroundings to our own interests, traits shared by both humans and wolves, reconcilable? Al Thieme, who recounted to me his experiences hunting elk in the home range of Oregon's Imnaha pack, reflected on this. "Nine months after our hunting trip, two of these wolves were killed by the state for their part in the death of livestock. During our trip, cattle were so plentiful on the public lands that we hunted on that it was customary to think you were hearing an elk, and then five minutes later you would come upon a cow wandering the canyons and meadows." Ever the philosopher, he went on, "Only when we as a culture can experience the ability to hold two contradictory concepts in our minds simultaneously about wolves, will we be able to allow them to thrive fully, and not limit the number of wolves in the ecosystem to serve the interests of a few individuals."

The contradiction: wolves are a biological competitor for resources that we desire, but their presence is also invaluable for creating conditions favorable to many other resources we also desire. Al's perspective, that wolves, a public resource, should not be killed for harming livestock, a private resource, so long as livestock are allowed to graze on public lands, is his attempt to find a course of action that reconciles this contradiction.

In the moment, it appears that our society has decided to attempt to embrace wolves, livestock, and hunting. As wolves spread into the Pacific Northwest landscapes where they have been absent for many decades, many other people will be asked to sort out how they wish to see these conflicts reconciled as well.

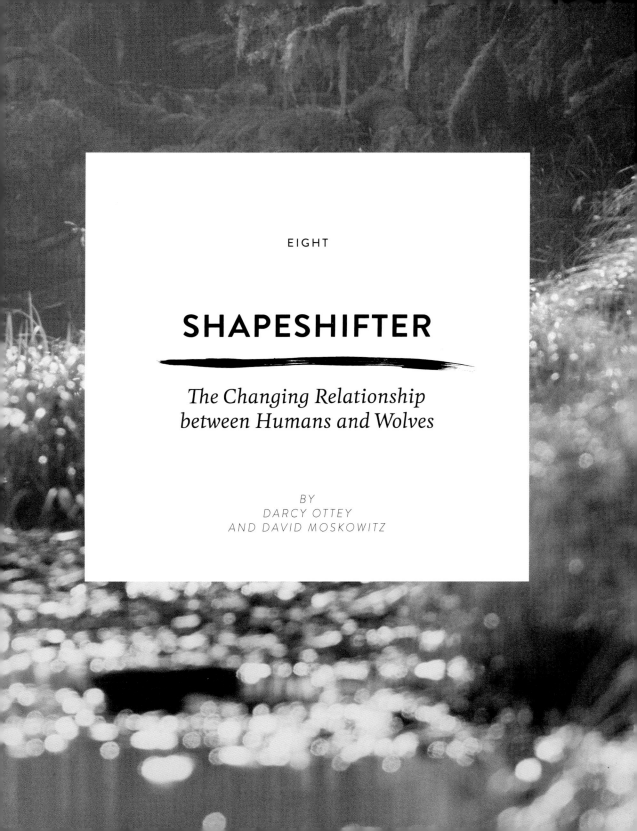

EIGHT

SHAPESHIFTER

*The Changing Relationship
between Humans and Wolves*

*BY
DARCY OTTEY
AND DAVID MOSKOWITZ*

TWO WOLVES popped out of the forest and trotted onto the beach just as the first rays of dawn brightened the day. A half hour earlier in the dark, my partner, Darcy Ottey, and I had traveled quickly along the ocean's edge where the surf would cover our tracks and scent, and then cut up to a rocky knoll on the edge of the beach to our makeshift blind. Darcy had joined me for the beginning of my trip to the area in the summer of 2011.

A light rain fell as the two wolves turned toward the ocean on the outer coast of an island in Clayoquot Sound. One of them sported a classic gray coat with reddish highlights. The other was primarily black. The gray one nosed a large strand of bull kelp on the water's edge, baiting its companion. A game of tug of war ensued, followed by a few races back and forth across the open sand. Eventually the gray wolf tired of the game, cut back to the water, and began trotting toward the end of the beach.

When it reached the point where we had left the water, it stopped and smelled, then followed our trail a few feet. It paused and looked up in our direction. We sat motionless in the blind. The wolf was no more than a hundred feet away. It continued its approach, nose down to the ground. The black wolf trotted over toward the gray one and smelled where it had started following our trail. The gray wolf approached closer, climbing up on the rock outcropping, now only twenty feet from us. Having not anticipated a wolf in that direction, we had constructed the blind fairly openly from that angle. We sat perfectly still in the dim light as the wolf approached closer, head cocked, mouth closed, and ears semi-erect. With these signs of both curiosity and trepidation, it took a step forward and then backed off a ways, then took a few steps forward again. It lifted its nose and sniffed intently, and finally stopped at about eight feet away. For a moment all three of us were perfectly still, wondering what was going to happen next.

The moment broke when the black wolf came up behind the gray wolf, drawing the other's attention. The two of them scampered off the rock and onto the beach. They circled around to the inland side of the small outcropping, smelling as they went, then paused again looking up at the blind. Another moment passed, and they turned and walked back down the beach, away from us, lying down in the driftwood on the forest's edge.

After the wolves moved off, Darcy and I chatted about what had just happened. I asked her if she had been scared. "A little nervous when it got so close," she replied.

PREVIOUS: Two wolves playing on the beach in a manner very reminiscent of dogs. Clayoquot Sound.

She asked me the same question, and I reflected. I just had two carnivores with the capacity to kill animals many times my size standing only a few feet away, and honestly I hadn't been scared at all. Excited, thrilled actually, but not scared. Why not? If a bear had approached me this closely I am sure both my body and mind would have been ramping up for a negative encounter. I would have had my bear spray in hand, ready to go. In grizzly country, if I even just get a sense that a bear might be about I may peel the Velcro to make the canister quicker to grab. Here, while I was aware of the bear spray clipped to my pack by my side, I never once felt the urge to have it in hand. To me, the wolves had appeared curious and skittish, not aggressive or predatory. Perhaps my reaction was to their body language.

For the next several days we observed these two wolves along with three more of their pack mates. They played and loafed on the beach, intermittently feeding on a rotting elephant seal carcass that was slowly sinking into the sand. Over those days the two wolves from the first day came toward the blind once more to check it out. The others appeared not to be aware of our presence. Watching the pack play, eat, and sleep, we could easily imagine somebody's family dogs on a vacation at the beach.

As we sat in the blind observing, Darcy and I grew curious about our reactions to the wolves. We wondered about the various ways that people think about wolves and relate to them in the Pacific Northwest. Our shared curiosity led us to write this chapter together, applying Darcy's background in the social sciences with mine in the biological sciences in our attempt to make sense of the way people relate to wolves.

Among the first people we talked to was Maquinna (Lewis George), hereditary chief of the Ahousaht First Nation. Ahousaht is the largest nation of the Nuu-cha-nulth people, one of three bands who have lived on Vancouver Island's Clayoquot Sound for countless generations. Maquinna, which is the name for the head-ranking hereditary chief, was gracious enough to speak with us in the middle of busy day in June 2011 in Tofino. "Wolves are really a sacred being for us," he explained:

> In our belief the wolf transforms into a killer whale. We feel that the wolf and killer whale are one in the same creature. We had a story from one of our villages: not too long ago, we had an elder that, she might of been about 110 when she passed away. Her name was K'wina. The reason why I know this story is because my mom passed away, and my dad remarried and that was her family. She frequently visited my stepmother and my dad. She was telling a story one day. She was saying that she was sitting on her porch in Nuchalitz. It's kind of like a bay, I've never been there but this is how she was describing it, it's a tiny little bay and she said she seen the wolves come in. The wolves came in, I mean the killer whales came in, and they went into this harbor and there was no way out, and it's quite a long ways for them to go. They have to come up to breathe in order to get in or out ... she saw them come in, and they never came back out. She

looked over on the beach, and she noticed all of these wolves that were all wet, coming out of the water. She actually seen it, the transformation. This would've happened maybe about seventy years ago, when she seen this. She was a young [woman]; maybe she would've been in the latter part of her thirties when that happened. So, you know, take it as how you want to take it, but that's what she saw. That's what she saw. So the killer whale and our wolf is one in the same being.

For many generations and across cultures, people have understood wolves as creatures who transform, inspiring awe, fear, and other emotions. Whether we take these stories as allegory or literal truth, wolves most definitely do shift. Their intelligence and adaptability allow them to shift their survival strategies to fit the specific environment they find themselves in. Wolves also shift the manner in which they interact with humans depending on the signals they receive from us. And certainly our perceptions of wolves shape-shift and transform, especially when differing cultural views collide.

Numerous researchers and philosophers have tried to explain the intensity of people's feelings about wolves: why some people hate them, why some people love them, and even why some people feel ambivalent toward them. But there's also another layer, as these various perceptions in turn influence public policy toward wolf management. Through the reams of literature on the topic, one very important theme emerges: there is often a disconnection between the physical wolf we may encounter in the forest, *Canis lupus*, and the wolf in our imagination. The "symbolic wolf," as a number of scholars, including veterinary anthropologist Elizabeth Lawrence refer to it, is a cultural construction—an idea of what a wolf is, based on symbols, language, and myths. The symbolic wolf may be communicated through stories passed down in families or through websites, movies, and other media of modern culture. Because we understand wolves largely through a cultural lens, as the dominant cultural forces in the Pacific Northwest have shifted, so too have the ways we understand wolves.

This shape-shifting symbolic wolf is of particular importance in our region for three reasons. First, at this point most people in the Pacific Northwest have limited personal experience with flesh-and-blood wolves. Second, the very existence of wolves in our region today is tied to politics, in that changing perceptions of their value have led to changes in management policies toward them. Third, many people imbue the symbolic wolf with meanings connected to their own deepest values and beliefs, such as the value of personal connection to nature, the value of freedom, or even how they feel about the role of government in our lives. When politics, values, and lack of personal experience with flesh-and-blood wolves come together, we rely on cultural constructions of wolves (often historical or imported from other regions) to make sense of the animal, and the actual physical wolf can become lost in the conversation.

How do we integrate this symbolic wolf with the real animals who are returning? To start with, we need to understand more about the various guises symbolic wolves have in the imaginations of Pacific Northwesterners and where those guises come from. Then we will consider how this understanding might affect our interactions with the flesh-and-blood wolves we now share the region with.

TRADITIONAL PERSPECTIVES ON WOLVES

First Nations peoples of the Pacific Northwest each have their own unique perspectives and cultural practices concerning wolves, an animal that has figured prominently in the cultural and spiritual lives of many of these diverse groups.

The Nez Perce have partnered with the USFWS in managing wolf recovery efforts in Idaho. As Jeremy FiveCrows has written, "The wolf has always been a symbol of strength, hunting prowess, and power." Elder Horace Axtell said that the Nez Perce "feel physically and spiritually connected with the wolf." First Nations' associations with wolves were not universally positive, however; the Chilcotin of what is now the interior of British Columbia are reported to have traditionally feared wolves, believing that "contact with the animal … cause[d] nervous illness and possibly death."

Joe Martin, Tla-o-quia-aht master canoe carver and indigenous rights advocate, told us that for the Nuu-cha-nulth, the wolf crest or symbol is among "the most important of the animals." You can tell this because they're frequently at the bottom of the totem pole, which is not the least important, as many people think, but is the base, because they "uphold the natural law." Wolves, bears, and killer whales are the animals commonly depicted at the base of totem poles. Both Martin and Maquinna (Lewis George) shared that traditionally among the Nuu-cha-nulth it was taboo to kill a wolf. However, other tribes did not have this taboo. For example, there are historical accounts of wolves being hunted by the Coast Salish and Quinault, First Nations of Puget Sound and the Olympic Peninsula, respectively.

The wolf has been central to the cultural life of First Nations up and down the Northwest coast through the practice of a ceremony that anthropologists commonly call the Kluukwalle or Wolf Ritual, also known as the winter ceremonial. It is unclear how widely the Wolf Ritual continues to be practiced, if at all, but the ceremony still has significant cultural importance to at least some First Nations, and many groups historically considered it to be among the most special, sacred, and secret initiatory ceremonies. Shrouded in mystery, the ceremony's name and origin stories varied by tribe, as did the ritual's elements and possibly its purposes. The origin stories often highlight the bravery associated with wolves. Indeed, these qualities are at the heart of the wolf ceremony, as anthropologist Alice Ernst wrote: "The wolf is the bravest of any animal in the woods …. They don't fear anything …. That is why the wolf is chosen."

According to Ernst's accounts, throughout the Northwest coast the Wolf Ritual was a multi-day wintertime event, typically held before the full moon. Villagers were invited to a potlatch but not told that it was this particular ceremony. Members of the Wolf Society (often considered the most secretive and prestigious of societies, the hunters), dressed as wolves, captured the young initiates and took them to the Wolf House, where they received training and instruction. This was a challenging and physically taxing ceremony. After days of training, the young people returned to the village and participated in another series of rituals. At the end of all of this, they were reborn into the tribe, once again human, but carrying the power of the wolf.

While times have changed, the wolf continues to have a significant place in the cultural and spiritual life for many First Nations, even in places with few if any physical wolves. This is certainly true of the Quileute Nation of the Olympic Peninsula in Washington, whose origin story tells of how the Quileute people came from wolves who were transformed into humans by a shape-shifting Changer named K'wati. As part of our research, we attended the Quileute Nations' Welcome the Whales ceremony in April 2011. This now-annual public ceremony, put on by the Quileute Tribal School, was resurrected in 2007. An important part of the event was a dance depicting the transformation of the whale into the wolf, the dancers wearing large, ornate masks and long cloaks, and moving in skilled and practiced wolf-like movements. Whenever wolf dancers entered into a dance throughout the day, the crowd hushed and focused attentively.

What accounts for the importance of wolves in the cultural lives of many of the First Nations in the Northwest? When we asked Maquinna (Lewis George), "What's special about wolves? Why do we as humans have a different relationship with them than other large carnivores, like bears and cougars?" he didn't even pause to think. His response was right there on the tip of his tongue: "None of the other animals that I know of have the ability to transform. That ability is very special and unique, and it's healing."

A very different symbolic wolf came to the Pacific Northwest with the arrival of European Americans: a symbol of evil, the icon of a cruel and heartless killer. This potent cultural construction originated in the agricultural societies of Europe and persisted in European-American culture, becoming an important factor in the extermination campaign that pushed wolves out of much of the Pacific Northwest during the 1800s and 1900s.

This perspective on wolves evolved over many centuries in Europe. Historically wolves were prominent and often positive symbols in European mythology and cultural history and in western civilization in general. Wolves are key in the story of the founding of Rome, for example, where the two brothers Romulus and Remus were abandoned by their human family and raised by wolves before creating the city. In this origin story of a society that venerated warriors, wolves were depicted as nurturing the founders of the entire culture. European stories that pre-date the expansion of a livestock- and agrarian-based economy paint wolves in a similarly positive light.

The hatred and fear of wolves that became pervasive in Europe grew in part out of wolf predation on livestock. With the shift from hunting wild game to tending domestic animals, people had a great deal more energy invested in specific animals as well as a sense of ownership over them. In a time before supermarkets or centralized government, the destruction of a flock of sheep by wolves might mean the threat of starvation for a family. People began to see the wolf as a conniving thief who comes in the night to rob humans of their possessions and threaten their survival, rather than as an inspirational hunter.

While depredation of domestic livestock spurred hatred and fear of wolves among pastoralists—groups who raise hoofed animals as livestock—it was probably wolf attacks on humans in Europe that grew this symbol of evil to its eventual mythic proportions. Wolf attacks on humans have been exceptionally rare in North American history, but a comprehensive international review documents a long history of sporadic but clustered incidents in Europe and Asia. Outbreaks of rabies in wolves accounted for many of these. There was no vaccine to protect people against rabies until 1885, and as is still the case, there was no cure. Indeed, the image of wolves as crazed killers is perhaps not too far off from the reality of a rabid wolf randomly and fatally attacking people. Other European accounts of wolf attacks on humans appear to have misidentified attacks by dogs or wolf-dog hybrids. Finally, there are accounts of wolves consuming the remains of humans who had died during disease outbreaks or on battlefields, both times when the ability to tend to the bodies of the deceased was overwhelmed by the quantity of corpses. Seeing wolves consuming the remains of humans was likely very disturbing to already traumatized survivors of such situations.

In Europe, fear and hatred of wolves seems to have gained the most traction during the Middle Ages, when the powerful Christian church demonized wolves as symbols of evil. Stories of werewolves became common, leading to mass killings of people, as when a single French magistrate between 1598 and 1600 sentenced to death six hundred citizens accused of being werewolves. In the wake of widespread extermination efforts, the last actual wolf was reportedly killed in England in the early sixteenth century, and throughout most of Europe wolves had either been exterminated or reduced by 1700 to what author Bruce Hampton called "remnant populations."

European colonizers carried this symbolic wolf to the New World, where they continued the battle against wolves that their ancestors had fought for centuries. The focus of their hatred was indeed already a mythical wolf—after all, wolves had largely been eliminated from their homelands before many of these Europeans came to North America. This is pretty spectacular if you think about it. The image of the bloodthirsty killer wolf persists, to some extent, in the imagination of many contemporary Pacific Northwesterners through stories, language, and cultural symbols, which came to the region from colonizers over the last two centuries, who carried this image from Europe, either directly or through ancestry, where in many areas there hadn't been wolves in generations.

Wolf biologist Luigi Boitani looked at cultural attitudes toward wolves in Europe, Asia, and North America, and concluded that "early human ecology types" influence later perspectives. The important and positive role that wolves play in Pacific Northwest coastal tribes fits a pattern that Boitani found in cultures around the world when people lived with wolves. In other regions where humans related to their environment primarily as hunters and warriors, the most favorable attitudes toward wolves persisted as well. Here wolves served as striking role models in hunting prowess, strength, and ferocity.

Conversely, cultures relying on livestock tended to have the most negative stories about wolves. While changes in a region's material culture sometimes contributed to shifts in representations of wolves, at times older cultural views persisted even as economies and physical relationships to wolves shifted. In Italy, for example, where early associations with wolves were positive, tolerance for wolves continued even as pastoralism and negative views from the north became a dominant force in their culture. Boitani suggests that early positive views of wolves might have contributed to their never having been extirpated from Italy, where a small population still exists today. In modern times in the Pacific Northwest, we can see something similar transpiring as historical cultural views on wolves influence contemporary views, even as new symbolic wolves come to the fore.

CONTEMPORARY PERSPECTIVES ON WOLVES

Even as traditional perspectives on wolves persist in parts of the Pacific Northwest, new symbolic wolves are emerging, reflecting changes in the material culture, in human relationships with the natural world, and in the intermingling of traditional cultural perspectives in the Pacific Northwest. These new symbolic wolves strongly influence the political landscape and direct encounters with wolves in the region.

Washington is the most culturally diverse state in the region, but even here there are six and a half times more European Americans than the next largest ethnic group, Latinos, and European-American cultural influence almost exclusively fuels mainstream political and cultural discourse regarding wolves, with the occasional exception of indigenous perspectives. First Nations' assertion of their preexisting rights to traditional homelands as well as to traditional hunting, fishing, and plant-collecting locations has led to a place at the table for them in some, though certainly not all, land and wildlife management decisions. Consequently these days the significant cultural and political divide on the subject of wolves in the Pacific Northwest is better described as socio-economic and geographic. It is also often described as an east-west cultural divide (referring to the divide created by the Cascade Range), but this is only partially accurate. In reality, it's more of an urban-rural split. Politically, culturally, and economically there are significant differences between the metropolitan regions of Vancouver, Seattle, Portland, and other smaller cities, and the more rural areas.

In his studies of attitudes toward wildlife in the late 1990s, Stephen Kellert found that urban-rural variations are a fundamental factor in perceptions of wildlife, though it's not as much about the population density of an area as it is about the "extent of a dependence on the land for a livelihood." Specifically regarding wolves, Kellert identified farmers, rural residents, and people with less formal education as more likely to view wolves as "threatening, unworthy, and sometimes evil," while younger, college-educated, and urban people tended to regard them more as "noble and admirable creatures possessing great moral and naturalistic significance."

I ran into Dan Warnock, a cattleman by trade, as he was wrapping up a day of range-riding during the summer of 2011 in the Wallowa Mountains. Attempting to reduce depredation on livestock by wolves, a group of local ranchers with funding and support from Defenders of Wildlife hired Warnock. His job was to haze wolves if they were close to livestock during the summer months when cattle are run through national forest allotments. Each day the ODFW would send a text message to Warnock and local ranchers with the general location of the GPS-collared wolves in the area, and Warnock would check things out.

We chatted as he loaded his four-wheeler onto a trailer behind his pickup truck. I had read about a variety of proactive measures that were being tried to prevent or reduce livestock depredations by wolves. Seeing this effort as an attempt to get beyond the typical divide between ranching interests and wolf conservation, I was quite surprised by one of the first things Warnock declared to me, "I don't like wolves . . . I do this job for the stockmen." Rather than looking for a way for Oregon ranchers and wolves to coexist, Warnock shared the perspective that he and ranchers in the area were doing the best they can with a deck that they feel has been stacked against them.

When I asked what he actually does while traveling around the forest, he responded that he is pulled in two directions. Defenders of Wildlife wants him to keep wolves away from livestock. This is hard, he noted, since cattle are everywhere out there. If he hazes wolves away from one cattle allotment, they might just go over into the next area, which also has livestock. The ranchers want him out there trying to document livestock killed by wolves so that there is a more accurate count of depredation events. Such documentation, along with accurate numbers, is also a prerequisite for ranchers to be reimbursed by the state for lost livestock, and in order to move forward the process of getting a "lethal take" permit from the state for wolves that are chronically depredating on livestock in accordance with the state's management plan. He expressed a great deal of frustration that the state wouldn't declare more livestock losses as confirmed wolf kills.

Given his clear distaste for wolves, I was curious about how he reconciled this with being paid through an organization that has been at the forefront of wolf recovery in the West for decades. When I broached the subject, he replied, "Defenders pats themselves on the back for funding this position . . . . But you pay for my position. Defenders gets federal tax dollars."

When I clarified this point with Suzanne Stone, northern Rockies representative of Defenders of Wildlife, she stated that all funding for the range rider position comes from private sources. Both perspectives are accurate to some extent: Defenders of Wildlife does get some federal funding, so tax dollars do fund some of their work. However, they also segregate their funds by source and allocate them accordingly. More interesting, though, is how individual perspectives influence the telling of what seem like objective facts and just how politicized even basic information is when it comes to wolves.

While the image of wolves as an evil and perhaps supernatural entity has mellowed, they remain a threatening force in the perceptions of many and, in many rural communities in the Pacific Northwest, a symbol of unwanted government interference in people's lives and in rural economies based on agriculture, livestock, or resource extraction.

With the listing of wolves as endangered under the United States Endangered Species Act in 1973, the wolf issue became even more political, and the federal government's decision to reintroduce wolves into the region in the mid 1990s further fueled the perceived connection between wolves and government interference. In certain areas, particularly in Idaho, reintroduction was very controversial politically. The topic remains fraught with frustration and anger, with some perceiving the decisions as being made by progressives who run things in Washington, D.C., or in the state government capitols west of the Cascade Crest in Washington and Oregon. In this view wolves connote urban liberals with little actual experience in the outdoors but a grandiose commitment to some idea of nature. As George McCormick wrote in an editorial regarding the formal decision of the city of Forks, Washington, to oppose the reintroduction of wolves to Olympic National Park, when this was being considered in the 1990s, "Some, who say the city council should have waited for scientific facts before taking a stand, are obviously not aware of the experience of old-timers in the area who lived with wolves. Who knows better the effect of wolves, those here who have lived with them or some city folks with letters behind their names whose only experience with wolves is in a zoo?"

Joel Reid provides a different sense of what wolves mean to many other contemporary Northwesterners. Reid, like Warnock, makes his living through work in wolf country, but in a very different capacity. Reid is an instructor at Outward Bound, an outdoor education program that teaches thousands of people nationally how to live and travel in the wilderness. He shared this account of seeing wolves in the North Cascades:

I woke up early, well before my students or co-instructors, feeling the draw of an early morning in the mountains. Looking out toward a beautiful Cascade peak, in the dim light of the new day, I immediately saw two objects moving on a slope across the valley. Their size, shape, color and the way they moved told me I was observing canines. After a moment I realized that not only that, they were wolves!

It was day five of a fourteen-day mountaineering course with a group of sixteen-to eighteen-year-olds from around the country. After a very long, tiring day of bushwacking and route-finding we chose a camp high in the peak drainage. We had decided to sleep in, but I was already in the habit of waking early. I had heard of wolf sightings in the Twisp River area with the return of wolves to the region, but I hadn't heard any stories yet first-hand. I am always excited by the chance to catch a glimpse of other large creatures traveling around in the mountains, but know that twelve pairs of boots crashing through the brush scares off most wildlife pretty quick.

Sitting up, still in my sleeping bag, I just watched for a moment. With graceful movements, they appeared to dance as they made their way down the hill. I got out of my sleeping bag and scrambled down-slope a bit to get a better view. There were definitely two wolves out there, one larger than the other, perhaps a mother and young?

As I stood there watching I wondered if they were aware of my presence. I was the only one awake so camp was quiet and still. As I scrambled down to a better viewpoint I accidentally kicked a small rock that tumbled down about twenty feet. They stopped instantly. They were now staring in my direction. I stood, mesmerized, witnessing what I knew was a rare sight in the North Cascades. I was so excited that I neglected to wake up any of my students or my co-instructor, Ryan, who reminds me of this fact still. When he did wake up and I told him what I saw, he jumped up but they had just dropped down valley and out of sight.

Watching them move across the slopes with such ease reminded me that no matter how much time I spend in the wilderness, and how good of shape I'm in, I'll never have that agility or grace. My leather boots tramping over rocks carrying me and my heavy pack will never quite fit into this environment. I'm just a visitor. But that's okay with me, with this chance to put things in perspective.

Reid had encountered another symbolic wolf, the icon of the wild. Exceptionally powerful since the middle of the twentieth century, this symbolic wolf has been a primary driver of wolf conservation in North America. For those who see wolves as a symbol of the wilderness and particularly those who live or work in urban areas, wolf recovery and conservation efforts are important attempts toward regaining the wildness felt lacking in the modern world. Where the wolf was once vilified for its association with wildness, from this cultural perspective wildness is the very reason that wolves are appreciated. The mountains of the Pacific Northwest can feel wild and pristine. Yet in many areas historical predators like grizzlies and wolves have been absent or highly elusive for several generations. The more travelers like Reid get to know the mountains, the more they are aware of this historic absence.

Seeing the wolf as a symbol of the wild also reflects an increasing appreciation of some traditional indigenous cultural values in the Pacific Northwest, a spreading understanding of the inherent value of nature and of the rights of other creatures. However, this modern cultural construction distorts the common traditional perspectives of Pacific Northwest First Nations. The dichotomy that separates people and nature (tame and wild) finds its primary origins in western civilization. From this dichotomy emerged a concept of wilderness, defined in federal legislation as a place "untrammeled by man, where man himself is a visitor who does not remain." Considering wolves as a source of inspiration may reflect a swing in where people place value between the poles of wild and tame, and the resurgence of the species similarly represents the influence of this shift in values on human political systems in the region. Thus, we value wild nature, but we still have written into legislation our separation from it.

MIRROR, MIRROR, ON THE WALL, WHO'S THE FAIREST WOLF OF ALL?

This dichotomy aside, a striking thing about the literature on wolves, whether in traditional stories, biological research, historical accounts, or elsewhere, is the consistent theme: humans and wolves are similar in many ways. From an ecological perspective, as an adaptable social carnivore, wolves are among the most similar species to humans in the Pacific Northwest. From a psychological perspective, the conscious, subconscious, cultural, and biological connection between humans and wolves makes wolves an excellent mirror for humans, an important factor in the exceptionally strong feelings they often elicit. As Eli Enns, Tla-o-qui-aht political scientist, said:

> I understand that there is a natural affinity among the Kakawin (Orca), Quiat-seek (wolves), and Kuu-us (real people). An elder from my family taught me that

it was Nuu-mock (immoral/taboo) for Kuu-us to kill or disrespect a member from the Kakawin or Quiatseek families without due and just cause. The affinity can be understood as a shared social structure; Kakawin live in pods, Quiatseek live in packs, and Kuu-us live and organize themselves in families. It is believed that all three beings feel and share love with one another, and would miss a loved one if they were lost. This is an important teaching because it fosters a better understanding of our own nature and our connection with the natural world.

While Enns's story reflects a conscious awareness of the similarities between wolves and humans, sometimes our connections between ourselves and wolves are not as conscious.

Whenever we find a mirror for an unseen or unacknowledged part of ourselves, it tends to provoke a powerful emotional response. Reading Reid's reflections on his encounter with wolves in the North Cascades, we glimpse his yearning for a sense of belonging and connection to the world absent for many modern people. Here wolves mirror positive qualities to which we aspire. As many people long for wild places and wild wolves to inhabit them, contained within this is a dissatisfaction with their own relationship with their surroundings. Some psychologists further posit that our longing for the wild in nature is a striving to access wild parts within our own human nature. How much does affinity for wolves reflect unacknowledged wild urges in our own hearts and souls?

Besides projecting our inner landscape onto wolves in a positive way, we also have a history of projecting our shadow sides, those parts of ourselves that we repress because we hate or fear them. The association of violence, treachery, and murderous behavior with wolves can be seen as subconscious human attempts to divert attention from our own propensity for these behaviors. In the sixteenth-century European executions of people accused of being savage, violent were-wolves, humans were exhibiting exactly that which they were seeking to expunge: their utmost capacity for violent, savage behavior.

LEARNING TO COMMUNICATE

The consequence of all these cultural constructions and projections is that they can leave us blind, unable to adapt to the wolf that is really there. When we encounter a physical wolf in the wild, we are not encountering an agent of the government, an ambassador of wilderness, a murderous killer, nor a harmless gentle creature. We are encountering an intelligent and adaptable wild animal. Just as wolves are returning to a physical landscape that has changed drastically since their ancestors last roamed large sections of the Pacific Northwest, so too has the cultural environment shifted for humans. Multiple generations of humans have lived, worked, and recreated across the region with limited or no interactions with wolves.

Rebounding large carnivore populations paired with human populations with little or no experience living with these animals has created conservation challenges, including that in reality, at times, these animals can be dangerous to people.

Like many animals, wolves are driven by a mix of instinct, learned behavior, fear of injury or death, and the urge to survive and reproduce. They are social animals, with methods for communicating their intentions and state of being. Yet for many in our technologically oriented society, understanding how to observe and interpret the presence or behavior of animals in the wild—much less understand how to respond ourselves—is a lost skill.

Systematic studies of wolves in Canada, Alaska, and worldwide suggest that wolves are the least threatening large carnivore in our region and in North America in general, with far fewer aggressive encounters reported than with grizzly bears, black bears, or mountain lions. Wolf-dog hybrids, wolves raised in captivity, and domestic dogs have a greater propensity toward aggression toward humans than wild wolves do.

Wolves can occasionally pose a threat to human safety, however. They are, after all, large carnivores with the capacity to kill animals far larger than humans. Many people don't want to think about this, preferring to dismiss dangers from wolves as simply part of an unfounded "icon of evil" interpretation of the symbolic wolf. Refusing to acknowledge the real, potential hazard that wolves pose for humans assumes yet another symbolic wolf, the "harmless wolf," in the words of Valerius Geist, which puts both species in danger.

In July 2000 two wolves were killed as a result of an altercation with a sea kayaker on a beach on Vargas Island, a British Columbia provincial park in Clayoquot Sound situated in the traditional territory of the Ahousaht First Nation. In this well-documented incident, the sea kayaker, part of a large group camping on the island, was the only one in his group sleeping outside of a tent when he was injured by the wolf. Another paddler had joined him sleeping by the campfire but retreated into a tent when he woke up in the middle of the night with a wolf sitting on the end of his sleeping bag. The camper drove the wolf off with noisemakers.

About a half hour later, the man still sleeping by the fire awoke to a wolf dragging him by his sleeping bag. The man yelled and struggled, and the wolf released him for a moment but then approached him again, biting him on his upper shoulder encased in the sleeping bag. After about five minutes, the other members of the kayaker's party were able to chase the wolf off but not before the paddler sustained a serious laceration to his scalp, requiring medical attention. The next day, two wolves, a male and a female approximately fourteen months in age, were killed by provincial conservation officers.

This is a cautionary tale suggesting what can happen to wolves and to people when we respond to a symbolic wolf rather than to the wild animal before us. As wolves return to the Pacific Northwest, we can prevent this sort of incident from repeating itself if we simply learn some basic communication skills. Animal communication is a fascinating field, full of all sorts of controversy and debatable research interpretations; in fact, a lot is going on when animals communicate that we don't understand despite all the research on the topic. When wolves communicate, they are using a mixture of gestures, facial expression, gaze, vocalization, and olfactory messages. When we enter into wolf territory, we are also communicating with wolves, whether consciously or unconsciously. We need to understand how to read the social cues we are receiving from wolves and what we're communicating as well. You can imagine that wolves are trying their best (as their lives may depend on it) to understand the cues that humans are sending.

We can learn about the consequences of poor communication from the Vargas Island incident. In fact, during the late 1990s and early 2000s, the west coast of Vancouver Island experienced a spate of aggressive wolf encounters. These events led to the WildCoast Project, a multidisciplinary, long-running, community-driven research initiative that continues to look at the relationships between humans and large carnivores in the region. This project followed the complete extirpation of wolves on the island by the 1960s, then a reversal of policies that resulted in the return of wolves over the next thirty years.

During this period outdoor recreation on the west coast greatly increased, perhaps as people sought a renewed sense of connection with nature. They often came, though, with little or no experience with wolves and wildlife. Educational programs and management strategies in the region discouraged behaviors that would prolong interactions with large carnivores, such as feeding them directly or indirectly, but some visitors ignored these recommendations. Asked why, people primarily reported believing that "interacting with wildlife was part of their connection with nature," according to a report. Respondents also believed their one interaction would not harm the animal. These comments suggest that some visitors were communicating potentially counterproductive messages to the region's wolves, creating an ongoing challenge for those concerned with improving human and wolf safety.

In his extensive research into the history of threatening and aggressive behavior by wolves toward humans, Mark McNay separated such behavior into six categories.

The first, investigative search or approach, is not actually aggressive on the part of the wolf but can be perceived as threatening by humans. This behavior indicates curiosity or a lack of fear. It might include such things as exploring human campsites, approaching humans, or escorting behavior (traveling parallel to a human for some distance). These behaviors have been reported prior to aggressive behaviors in individual animals, including on Vancouver Island where wolves use landscapes also heavily used by people.

Agonistic behaviors are those associated with social hierarchy, dominance, or a perceived threat. In wolves, behaviors that assert or test dominance include growling, barking, or baring teeth while approaching a human.

SAFETY WITH WOLVES

WOLVES ARE FLEXIBLE IN THEIR BEHAVIOR AND IN HOW THEY INTERACT WITH HUMANS. HOWEVER, SOME GENERALITIES ARE GOOD TO CONSIDER IF YOU'RE TRAVELING IN WOLF COUNTRY.

• In most of the Pacific Northwest, wolves are generally wary of humans, and most encounters between the two species are short-lived as the wolf will attempt to remove itself from the situation. In fact, more interactions between humans and wolves happen without the human's awareness of the wolf, who detects the human and moves away.

• In landscapes where they are protected, such as national parks, wolves may have a higher tolerance for human presence. In places where they are hunted, they will likely be very wary. Because poaching of even officially protected populations of wolves is so pervasive, most wolves in the Pacific Northwest exhibit wary behavior.

• Wolves who inhabit areas with higher human population densities tend to have more tolerance for interactions with people, while wolves in more remote settings can be greatly disturbed by even nonconfrontational interactions with humans.

• Young wolves—pups and yearlings—are the most likely to demonstrate curiosity about humans. A curious wolf might approach a human for a better view or circle around to get

Encountering a wolf in the wild can be exciting. How a wolf will behave during a human encounter depends on the context, the wolf's prior experience with humans, and the cues the wolf receives.

downwind of the person to catch the scent. Similar to an encounter with an unleashed dog in the park, such behavior from a wolf can be either exciting or concerning, depending on the specific situation and the person's past experience. Here's a place where it's good for us to be intentional in how we communicate with wolves. As with dogs, using body language that projects a sense of calm and assertiveness is appropriate in these circumstances.

· It is important to be able to recognize a wolf's unusual, aggressive behavior to ensure that you protect yourself and to discourage any escalation of the behavior. Behaviors that might be cause for concern are just what you might associate with a poorly behaved, aggressive dog, including barking, raised hackles, teeth baring, charging, chasing, or stalking of people or pets traveling with people. If a wolf exhibits any of these behaviors, it is communicating that something's really wrong.

· In response to aggressive behavior, you want to communicate strong boundaries and that the wolf should leave you alone. The best responses are

similar to those for dealing with mountain lions or domestic dogs (or humans, for that matter, as Darcy learned in a human self-defense class). Stare at it, make loud noises, and stand at full height in an assertive posture. Send the message to the wolf that you are aware of its presence and not intimidated. If the wolf is testing you, responding in such a manner expresses a lack of fear, a deterrent of further aggressive behavior. Maintain eye contact with the wolf and avoid running, crouching, or turning your back on the wolf, as these actions could initiate an instinctual chase response.

· If, despite initial efforts to deter an aggressive wolf, it continues to approach or engage with you, throwing objects at it and hitting it with anything available are appropriate and generally very successful at deterring the animal. Either pick children up, or place yourself between them and the wolf. Pepper spray designed for bears is also an effective deterrent for wolves.

Predatory behavior, associated with hunting and attempts to subdue prey for the purpose of consumption, has never been documented toward humans in the Pacific Northwest, though one or possibly two cases have now been reported elsewhere in North America.

Charging behavior is associated with either prey-testing or dominance behavior, but distinguishing between the two can be impossible in the moment. In a charge, a wolf runs toward a human either to threaten or attempt to get the

A wolf's gaze can be riveting. What we see in the eyes of a wolf says as much about ourselves as it does about the wolf.

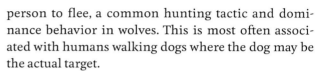

person to flee, a common hunting tactic and dominance behavior in wolves. This is most often associated with humans walking dogs where the dog may be the actual target.

Self- or conspecific defense is aggression associated with trapped or injured animals, or associated with humans approaching pups, den sites, or rendezvous locations. This often includes barking but rarely leads to physical encounters initiated by the wolf.

The final category is the behavior of diseased wolves. Animals infected with rabies show aggressive and unpredictable behavior often associated with impaired motor skills. There are no historic or modern records of rabies in wolves regionally. In North America rabies in wolves is almost exclusively associated with areas where they live alongside Arctic foxes (*Vulpes lagopus*), the primary carrier of the disease, a species that does not occur in the Pacific Northwest.

McNay's study concluded that "the threat of unprovoked aggression from nonhabituated wolves is very small." His research showed that wolves habituated to the presence of humans account for the majority of aggressive encounters, most of which involved wolves conditioned to associate humans with food. Secondarily, in the absence of effective communication from humans, wolves engaging in dominance behaviors became aggressive toward humans.

A habituated animal becomes more tolerant of close association with humans because when they are around humans, nothing bad happens. In other words, people don't communicate to wolves that humans are dangerous for them. An increase in wolf numbers and an increase in the number of humans treating wolves benignly while traveling in wolf habitat combine to create excellent conditions for habituation. However, habituation is not necessarily dangerous in all its forms. Wolves who don't flee immediately at the sight of humans can allow our species to observe and understand them better. Habituation has been used as an exceptionally effective research technique and has produced some fascinating studies of wolf biology and behavior. Unfortunately for wolves, feeling comfortable in the presence of humans can be dangerous. Wolves that are habituated to humans can be more susceptible to legal or illegal hunting as well as to trapping and poisoning.

Habituation typically leads wolves to associate humans with food (not that humans are food, but that humans have food) in several ways. Humans may feed wolves directly by placing food out for them, usually to prolong the interaction or create an increased sense of intimacy. Humans may indirectly feed wolves when we don't clean up after ourselves, leaving food out in a campsite or leaving garbage unsecured. Wolves are curious and adaptable in their diet and can therefore be attracted to food odors in human areas. The line "a fed bear is a dead bear" refers to unsavory encounters that result from food conditioning of bears, which are far more prone to food conditioning than wolves, perhaps because of their even more adaptable omnivorous diet and foraging behaviors. Conditioning wolves to associate human presence not with danger but instead with food can rapidly lead to aggressive behavior. Food conditioning appears to be a central part of what happened in the incident on Vargas Island. There had been multiple accounts of these wolves becoming more assertive in the months and weeks leading up to the incident as well as reports of people feeding them on this beach. A review of all the negative encounters between wolves and people from the west coast of Vancouver Island found a similar pattern in many of them. Poor—and likely unconscious—communication choices on the part of humans were the starting point.

The social and ecological conditions that led to this aggressive behavior could easily recur in other parts of the region as wolves expand their range. However, the situation is totally preventable with some good housekeeping. In Yellowstone National Park, where measures to prevent food habituation of bears have been in place for decades, food conditioning of wolves has not been an issue despite millions of visitors and wolves that are accustomed to their presence. Camping behaviors that minimize interactions with bears also work with wolves: keeping a clean camp, cooking away from sleeping areas, and securing food storage, in other words, communicating to wolves that humans are not a source of food.

When humans bring dogs into places where wolves live, they are adding another dynamic into the complex interspecies conversation. Wolves and dogs have a long history as domestic dogs are the descendants of habituated and food-conditioned wolves. Interactions between wolves and dogs are highly variable, including antagonistic, predatory, amiable, and even interbreeding at times. Attacks on dogs both on and off leash, and behavior where wolves "escort" people and dogs, have been documented in various places in the Pacific Northwest. In habituated populations of wolves, aggressive behavior toward dogs in the presence of humans is not atypical. The dynamic involved in aggression toward dogs may be similar to that of wolves interacting with closely related coyotes, a typical behavior pattern between competitive carnivores. Behavior used to address threatening wolves when a dog is not present is appropriate in these situations as well.

On the other side of Vancouver Island from Clayoquot Sound, the northeastern coast is separated from the British Columbia mainland by Desolation Sound. I traveled to the area to meet with Sabina Mense, who manages a citizen wolf-monitoring project on Cortes Island. Desolation Sound, a glacier-carved trough now flooded by the Pacific Ocean, is home to another scattered archipelago of forested and lightly inhabited islands. The current resurgent wolf population of Vancouver Island most likely originated from animals who made their way from the mainland, island-hopping through this area. On my drive and ferry ride over, I was struck that the islands are similar in feel to the more populated San Juan Islands which span the northern portion of Puget Sound between Vancouver Island and Washington State.

On a sunny July day, Mense took me out to a previous rendezvous location of Cortes Island's resident wolf pack. Perched on a bluff that drops off steeply toward the sound below, covered with manzanita and madrone, this was an entirely different environment than I had ever seen for a wolf rendezvous location. From a vantage on the edge of the rocky bluff, she told me about how she had discovered the area and its use by wolves. A year earlier she was conducting a survey for a conservation easement and had scrambled up the side of the bluff. Close to the top, she saw some movement and realized that a wolf was looking down at her. The wolf "woofed" at her. "The wolves hazed me," she said. "That was their front yard." She backed down the slope and left the area.

This story beautifully illustrates what it looks like when someone is paying attention in wolf country and knows how to communicate respectfully with the animals. Mense got an exhilarating experience with wolves that day, and the wolf got some privacy for its pups. How many other people, not understanding what the wolf was trying to say, would have kept moving toward the wolf, unknowingly communicating a threat? As Mense walked up to that bluff, no doubt she carried her own sense of the symbolic wolf. But equally as important, she saw (and heard) the physical wolf and paid attention to what it had to say.

It's not that we want to get rid of the symbolic wolf. As that wolf shifts and changes, it helps us, giving us insight into our fears, our sense of awe and inspiration, and our shadowy depths, whether individual or cultural. Symbols are important. Meanwhile, the physical wolf has its own right to exist for itself, for its place in the family of beings. Its continued existence enriches humans, feeding the symbolic wolf as our cultures shift and change. And, of course, some would dispute that there is any difference between the physical and symbolic wolf. Wolves after all are shapeshifters. Who are we to say where the killer whale ends and the wolf begins; where the symbolic wolf ends and the physical one begins; where we as humans end and wolves begin?

NINE

ISOLATION

*Lessons from the
Olympic Peninsula and Beyond*

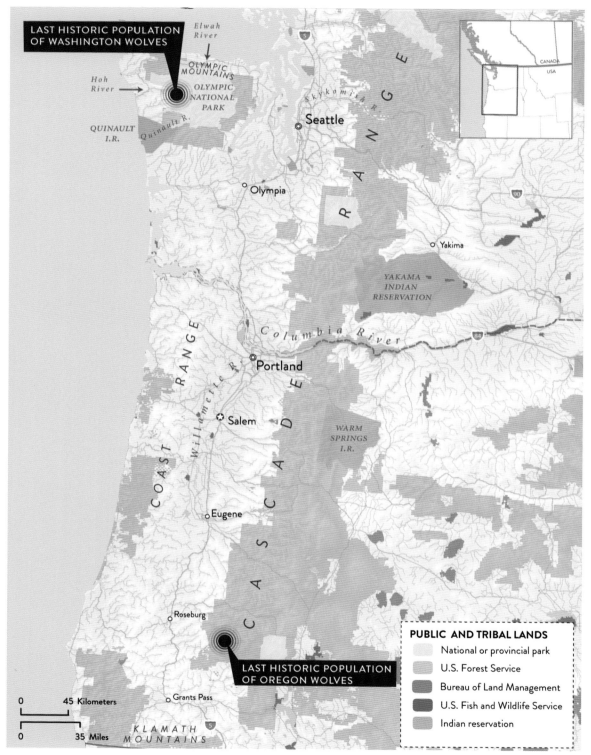

LAST HISTORIC POPULATION
OF WASHINGTON WOLVES

Elwah River

OLYMPIC
MOUNTAINS

Hoh River →

OLYMPIC
NATIONAL
PARK

Skykomish R.

QUINAULT
I.R.

Quinault R.

Seattle

Olympia

Yakima

YAKAMA
INDIAN
RESERVATION

Columbia River

CANADA
USA

COAST

RANGE

Willamette R.

Portland

Salem

WARM
SPRINGS
I.R.

C A S C A D E

R A N G E

Eugene

Roseburg

LAST HISTORIC POPULATION
OF OREGON WOLVES

0   45 Kilometers

0   35 Miles

Grants Pass

KLAMATH
MOUNTAINS

PUBLIC AND TRIBAL LANDS
National or provincial park
U.S. Forest Service
Bureau of Land Management
U.S. Fish and Wildlife Service
Indian reservation

COASTAL WASHINGTON AND OREGON

AS ONE HEADS SOUTH on the Northwest coast, leaving behind the wilds of the Great Bear Rainforest, the character of the landscape changes, incrementally in terms of the climate but starkly in the human footprint on the land. West of the Cascades in Oregon and Washington, rather than continuous and mainly unroaded primeval forests, wildlands here are generally defined by massive clearcuts, tree plantations, eroded stream banks, decimated fish stocks, an extensive road system, and urban and rural development. However, sections of inaccessible mountains harboring wild rivers teeming with salmon beneath ancient trees still remain in some places.

Wolves clung tenaciously to existence in this part of the Pacific Northwest longer than elsewhere in the region. The last confirmation of a wolf specimen collected in Washington State was in the early 1920s, on the western side of the Olympic Mountains. The last bounty paid for a wolf killed in Oregon was in the southwest part of the state, in the Umpqua National Forest on the western slopes of the Cascades in 1946.

The Klamath and Siskiyou Mountains, which span the Oregon-California border, anchor the southern end of Oregon's Coast Range. This botanically and geologically diverse, jumbled complex of mountains is bisected by deeply cut wild river valleys. North of this, the Oregon Coast Range has largely been roaded and clearcut, and is primarily managed for timber production. While also heavily logged in many places, the west slope of the Oregon Cascades retains tracts of roadless wild forests that remain in similar conditions to when they acted as the last refuge for wolves in the state. Starting north of the Klamath Mountains, the Willamette Valley forms an enclave of agricultural lands and urban development that divides the Cascades from the Coast Range. The Columbia River Gorge slices through the mountain ranges, the only river cutting through the Cascades, defining the border of Oregon and Washington.

The southern portion of western Washington resembles western Oregon, but around the middle of the state things begin to change. Here the rolling Coast Range gives way to the rugged Olympic Peninsula which separates Puget Sound from the Pacific Ocean. During the most recent advance of continental ice, northern Washington was covered with ice. Ice carved out Puget Sound and sheared the sides of the Olympic Mountains. East of the Olympics, the lands around Puget Sound are heavily developed in a mix of urban, suburban, and agricultural sprawl

---

PREVIOUS: Dense rainforests blanket the Hoh River valley on the west side of the Olympic Mountains in Olympic National Park, Washington.

which spreads across the lowlands to the base of the Cascades and from the Columbia River in the south to Vancouver, British Columbia, in the north. The lower elevations of the Cascades are heavily roaded and logged, but higher elevations are protected in a series of wilderness areas and two national parks. Olympic National Park is the core of the largest wilderness area in coastal Washington.

THE DEMISE OF WOLVES ON THE OLYMPIC PENINSULA

On a reasonably clear day, the high peaks of the Olympic Mountains are visible in the west across Puget Sound from Seattle. Despite their seeming proximity to the region's largest metropolitan area, once you make it out to the peninsula you realize you are a long way from the transportation and commerce hub of the Northwest. Log trucks rumble by on the two-lane highway. Clearcuts and tightly planted tree plantations dominate the landscape around the park.

Eleven major rivers originate in the high peaks of the Olympics, radiating out from the center of the peninsula toward the ocean or Puget Sound. Most still support abundant runs of salmon. Even today, to reach the west coast of the peninsula from Seattle is a five-hour driving trek, starting with a ferry ride across Puget Sound and ending with miles of winding roads. If the peninsula still seems remote, it was a world unto itself in decades past.

The history of human relationships with wolves on the peninsula and of their ecological role in this unique landscape cuts across many of this book's themes. Wolves are central to many of the peninsula's First Nations cultures. Conversely, much of the modern European-American rural population here is generally antagonistic toward wolves. Finally, with its national park, the peninsula attracts millions of visitors from all over the country and the world, the majority of whom have strong conservation values and would likely support the reestablishment of wolves in this area. Finally, restoration of extirpated native species is part of the national park mandate, and wolves are the only native mammal species currently missing.

From an ecological perspective, the absence of wolves in the Olympics plays a part in several interesting stories unfolding on the landscape. The first involves elk, a primary prey for wolves. Another includes a novel predator's recent arrival to these mountains, the coyote, and that animal's impact on one of its prey species, the Olympic marmot.

The lush Olympic rainforests, with some of the largest trees in the world, were the final refuge for wolves in Washington State.

As in most places, the beginning of the end for wolves in the Olympics arrived with an influx of homesteaders and livestock onto the peninsula starting in the early 1890s. At the peak of livestock grazing activity, cattle and sheep were run in every river valley, including locations that are roadless and wild inside the national park today. With increasing livestock grazing, hunting, trapping, and poisoning of wolves ramped up. Bounties paid on predators were an important income source for some homesteaders and were further incentive for killing wolves and cougars. Dora Richmond, who grew up on a homestead in the Olympics in the early 1900s, recalled her childhood experience with wolves:

We would put out the strychnine at night and pick it up in the morning, so it wouldn't get so many daytime animals, birds especially. Anyway, my aunt and I went out in the morning and there on the sandbar was a big old wolf. He was dead but my aunt had to take a shot at it anyway. There was a dollar bounty on wolves then so we wanted to skin it out. But the poison ruined the skin so we just put it in a bucket, took it out and buried it. . . . I was about twelve when the last wolf got died. My dad poisoned they cause they ate the sheep.

As Charlie Anderson, whose family was among the original homesteaders on the Hoh River on the western side of the peninsula, succinctly put it: "There again people had a different attitude, in those days the land had to provide." This hand-to-mouth utilitarian perspective that saw livestock and elk as food and wolves as a threat to survival, led to not only the demise of wolves, but the near extinction of elk on the peninsula as well—hunted for subsistence as well as for meat to sell. In 1909 President Theodore Roosevelt created Olympic National Monument to protect the remaining Roosevelt elk, named after him. In 1938 President Franklin Roosevelt expanded the monument into the current Olympic National Park.

By the time the national monument was established wolves had already become scarce in the Olympics, limited to a few river valleys on the west side of the peninsula. When the United Biological Service sent Olaus Murie (who went on to become a champion of wilderness conservation in the twentieth century) to trap wolves in the Elwah River Valley, he found tracks and signs of wolves but was unable to trap a single animal. During the winter of 1916, his field notes included his discovery of tracks in the Press Valley in the upper reaches of the Elwah, now part of a popular backpacking route in the park: "A wolf has been walking around—he came to my old tracks and smelled them before crossing and going on." The valley probably looks very similar today to what Murie found while checking his traps.

How wolves ultimately disappeared from the Olympics remains somewhat shrouded in mystery. The last confirmed killing of wolves there was in 1920 when a pair was killed on the North Fork of the Quinault River. That same year, a single wolf was killed in the Elwah River valley. Following these definitive findings, highly reliable sightings cropped up through the 1920s and into the 1930s, with credible sightings even as late as the early 1950s.

What became of the wolves of the Olympics? In 1948, in what was the definitive reference on mammals in Washington State for many years, Walter Dalquest speculated, "I doubt that man killed them all; perhaps some introduced disease, such as rabies, brought about their extinction." Besides rabies (which has never been documented in wolves in the Pacific Northwest), another possibility that comes to mind in the salmon-rich rainforest of the Olympics is *Neorickettsia helminthoeca*, the bacterium that causes salmon poisoning. In their seminal book *The Wolves of North America*, written around the same time, Stanley Young and Edward Goldman discussed the possibility that salmon poisoning was a contributing factor to the decline of wolves in coastal Washington and Oregon and even cited a source who hypothesized that it might have contributed to the final disappearance of wolves from Oregon.

In any case the demise of Olympic wolves clearly occurred after a large portion of the peninsula was protected in a national monument, a classic example of the tenuous nature of small populations and their vulnerability to unpredictable natural events. Even if the small population of remaining wolves was safe from direct killing by humans, the limited size of the remnant population would have made it far more susceptible to extinction from a naturally occurring disease. The final nail in the coffin was the isolation of this remnant population from other wolves. Without an adjacent source to restock otherwise suitable habitat, the species disappeared. This inherent threat to small, disconnected populations is a modern conservation challenge and the source of concern about numerous populations of large carnivores around the world.

Interest in returning wolves to Olympic National Park began as early as 1935, when Adolf Murie (brother of Olaus and one of the most influential researchers from his time in the field of wolf ecology) floated the idea as part of his report on the condition of the Olympics' elk population.

An exhaustive feasibility study of returning wolves to the Olympics was commissioned by the USFWS in the late 1990s. After evaluating the biological, economic, and social ramifications of the return of this contentious species, the study's authors concluded that the peninsula could support a viable wolf population again. More recent broader analyses of potential wolf habitat in the Pacific Northwest all recognize the peninsula as having a large amount of quality habitat for the species as well.

At first it may seem odd that the last refuges for wolves in Oregon and Washington were west of the Cascades, closer to human population centers. However, these areas also include the densest forests in the region. Around the world wolves have demonstrated an ability to persist in forested environments long after humans have destroyed their populations in open terrain. Perhaps the most notorious response to this situation was in seventeenth-century Scotland, where all the forests were burned to the ground to rid the country of wolves once and for all.

A little closer to home in both space and time is the story of wolf population expansion in the Great Lakes region of Minnesota, Michigan, and Wisconsin, where a recent summary predicted that "Wolf populations in the Great Lakes region are not likely to expand much beyond the heavily forested areas currently occupied." However, before European-American settlement, wolf densities were likely highest in the southern portions of that region in the open hardwood savannahs where large populations of deer, elk, and bison provided far more prey than in the dense forests to the north. As land-clearing activities moved north, wolf

populations in these areas decreased as well. Northern Minnesota, heavily forested and largely inaccessible to people, was the only location in the lower forty-eight states to retain a breeding population of wolves continuously through the 1900s.

A Roosevelt elk peers through moss-covered branches in the rainforest. Olympic Mountains, Washington.

Since the invention of the modern rifle, the disadvantages wolves face in open terrain is fairly obvious. It's not hard to imagine wolves disappearing into the dense, dark rainforest where a human's field of view can be less than a hundred feet

in any direction. Humans are visual hunters, and forested environments hinder our ability to use our powerful eyesight to find our prey. The second challenge forested environments pose to humans trying to find and kill wolves is access, dense forests being much harder for humans to navigate than open terrain.

RAINFOREST ELK WITHOUT RAINFOREST WOLVES

On an overcast day in early April of 2010, I navigated a maze of creatively labeled logging roads outside Olympic National Park. Using a GPS unit and three different maps, none of which showed all the roads I encountered, I slowly made my way up toward the park boundary and the end of the road in the valley of the South Fork of the Hoh River. Entering clearcuts that provided a view to the east, I could see my destination, a definitive line in the forest—on my side, a mix of clearcuts and second growth thick with alders and young trees; on the other side, an unbroken dark sea of ancient trees disappearing into the low clouds hiding the ridge tops and mountains to the east. Finally, having avoided being flattened by a falling tree at an active logging operation shortly before the end of the road, I donned my backpack and set out into the forest just as a fine rain began.

The cold waters of the Hoh River originate in the deep snows and glacial ice on the highest peaks in the Olympic Mountains. Once the river descends the steep gradients at its headwaters, it widens and winds back and forth across the broad valley bottom, which itself was carved out by much larger ice age glaciers. Each spring, the melting of mountain snowpack raises the river to flood levels, carving out new channels, devouring riverbanks, sweeping into riparian forests, and depositing silt. The flood plain and riverbanks are flush with deciduous trees, big leaf maple, black cottonwood, and red alders which take advantage of the break in the dominating canopy of evergreen conifers that blanket these mountains. Walking through the ancient stands of giant Sitka spruce, western red-cedar, and western hemlock above the floodplain, it is very easy to feel very small. Moss and lichen drapes from tree limbs and leaning snags, soaking up the sound of the river as you wander away from it, leaving you alone in a sea of tree trunks and fallen logs.

The South Fork of the Hoh was the site of several studies that examined the ecology of Roosevelt elk in the rainforest. Similar to the project in the Blue Mountains, researchers used a series of fenced exclosures to study how the presence or absence of elk activity influences the structure and constituents of the forest understory. In 1980 researchers erected two exclosures, each about 1.25 acres, in the valley bottom old growth forest. Five years later, Patti Happe used the exclosures to measure the effects of deer and elk herbivory on the rainforest as part of her Ph.D. studies. Happe, now the chief wildlife biologist for the national park, gave me directions to get to the study plots, now over thirty years old. I followed a lightly used hiking trail along the river for several miles into the park, then headed off through the forest, picking my way around fallen logs until the fenced exclosure came into view.

Happe's research suggested that, similar to what happened in the Blue Mountains, deer and elk foraging decreased shrub cover and increased mosses and grasses on the forest floor. Since her research, the population of elk in the South Fork had dropped so that the differences were not as stark now as they had been in the 1980s when the elk population was higher. Even so, the contrast was still quite apparent. Inside the exclosure was a tangled thicket of shrubs. Large patches of open, moss-covered ground, interrupted by patches of shrubs, dominated the forest around much of the exclosure.

While wandering around the surrounding forest, I stumbled across the remains of a recently deceased cow elk. As I approached, I noted coyote tracks in a muddy spot on the elk trail leading into the small clearing where the elk lay. The carcass was almost complete, with the exception of a missing eye on the side of its head facing up, likely the work of a raven, and a hole in its rump, a typical starting point for feeding by coyotes. Coyotes killing an adult elk is almost unheard of, so the coyote most likely had discovered the elk after it had died. The elk had a radio collar on it. Its distended belly indicated the cow had probably been pregnant.

Another half mile up the trail, the second exclosure told a similar tale. Inside the exclosure's eight-foot-tall fences, salmonberry shrubs grew high, branching over the fences, while outside the fence were grass and stunted, two-foot-tall salmonberry clumps. Where the trail came close to the river, I dropped down to the floodplain. The tracks of elk and river otters cut across patches of sand. I crossed the river on a tangled log jam. On the far side I wandered through a stand of alders, noting numerous elk antler rubs on the smooth-barked trees, a sign this area is used well during the fall breeding season. The tracks of a black bear meandered along the riverbank. I followed it up into the forest. Poking around under the trees, I found piles of bones from several salmon, likely a remnant of a bear's foraging activity from the fall prior.

When I got home several days later and told Happe about the dead elk and collar, she explained that this was an elk she had collared and had been monitoring.

When she went to the site months later, she concluded that the elk had probably died of malnutrition.

Elk (*Cervus elaphus*) evolved in grassland landscapes, primarily grazing on grasses and forbs, traveling in large herds on open terrain. Like wolves, they are a circumboreal species—one that evolved in Eurasia and then migrated into North America. Roosevelt elk (*C. e. roosevelti*) are the largest subspecies of elk in North America, slightly larger than the Rocky Mountain elk (*C. e. nelsoni*) which inhabits the interior of the region. Only slightly smaller than moose, bulls weigh from six hundred to one thousand pounds and cows between four hundred and seven hundred pounds. Roosevelt elk now have a patchy but expanding distribution from the redwood forests and coastal meadows of northern California north to southwestern British Columbia, where they have adapted to the mountainous rainforests of the Pacific Northwest coast. The Olympic Peninsula is home to the largest single herd of this subspecies.

Roosevelt elk have two distinct home range strategies. Some are migratory, wandering up to subalpine forests and meadows for the summer, returning to the rainforest valleys for the winter to escape the heavy snowpack of higher elevations. Other groups are nonmigratory. These elk spend the entire year in relatively small home ranges that center on valley bottom habitat, often including natural clearings as key features. In the winter, migratory and nonmigratory elk may share the same lower elevation locations.

Before intensive homesteading on the peninsula in the 1890s, elk appear to have been abundant, but within just a decade of subsistence and market hunting by settlers, elk numbers plummeted. One estimate of elk numbers at their low point put the peninsula's population at two thousand animals. In response to this, in 1904 a bounty was placed on cougars (in addition to one on wolves), and in 1905 a moratorium on elk hunting was established, lasting until 1933. These actions appear to have reversed the declines, and by 1915 there were reports of an overpopulation of elk with significant impacts on the browse species on which they relied. Subsequently several severe winters led to significant elk die-offs. Population estimates at this time put the population at seven to eight thousand animals. When Olaus Murie surveyed the peninsula again in the mid 1930s, he came up with a similar estimate of population size. Concerns about deteriorating range condition persisted, and again in the mid 1930s severe winters led to significant die-offs of elk. The moratorium on elk hunting was lifted in 1933, but in 1938 the national park was established and hunting was stopped once and for all inside its boundaries.

Since then elk populations have been relatively stable on the peninsula, with continued occasional fluctuations from die-offs during harsh winters and

OPPOSITE ABOVE: This fence is part of an exclosure erected to study the impacts of elk browsing on forest structure. Before construction, areas inside and outside of the exclosure were identical. Outside of it, elk have prohibited brush and saplings from growing while inside the plants have flourished. Olympic Mountains, Washington.

OPPOSITE BELOW: This red huckleberry branch illustrates the stunted appearance caused by repetitive browsing by elk and deer on many shrubs in the Olympics.

BELOW: The open moss-covered forest floor in this part of the South Fork of the Hoh River is indicative of heavy use by elk which remove shrubs and ferns over time.

subsequent rebounds. In the mid 1980s, the WDFW estimated the elk population of the peninsula, not including animals in the national park, to be about twelve thousand. By 2000 the herd had dropped to about 8600 according to WDFW estimates. Animals inside the park likely account for about three thousand more individuals. Unfortunately, there is no documentation of the abundance of elk prior to the period of intensive colonization activities, nor of the condition of the forests then, related to elk and deer populations. We also don't know what sort of changes in First Nations' hunting and management activities occurred as a result of colonial influences and how this influenced ungulate populations.

Early accounts of homesteaders document an open grassy understory in sections of forest along the Olympic coast. These locations were often preferred sites for homesteading. Such parklike openings in Northwest rainforest require some sort of persistent disturbance to create and maintain. Elk and deer were an important food source for the First Nations west of the Cascades. Farther south in the region, in the Willamette Valley and Klamath-Siskiyou areas, First Nations people burned landscapes to maintain forest openings for the benefit of deer and elk which they hunted, but there are no records of similar traditional management activities carried out by the First Nations people of the Olympics. In fact it seems likely that the elk themselves modified parts of the landscape through their foraging activities.

Numerous studies have shown that long-term foraging pressure from hoofed mammals, including elk in the Olympics, can decrease forest understory, retard recruitment of young trees and shrubs, and eventually create open, parklike forests with well-spaced large trees and a primarily grass and forb understory. Such changes can improve the quality of forage for elk in some instances, though the lack of shrubs to browse can be disastrous during winters with deep snow at lower elevation that prohibits deer and elk from reaching grasses on the forest floor.

The lack of streamside vegetation along this section of the Hoh River is due to heavy browsing by elk. Removal of shrubs from along streams can increase stream-bank erosion, as pictured here, which can negatively impact habitat for fish and other aquatic species.

Interestingly, if elk were responsible for creating the coastal openings, it would suggest that wolves were not acting to significantly limit the population of elk on these parts of the peninsula.

Studies of wolves and prey populations in coastal British Columbia and in other forested environments have suggested that wolves do limit ungulate numbers in some instances in similar habitats to the Olympics. On parts of Vancouver Island, wolves live with both black-tailed deer and elk, and deer are the primary prey species. During a decline in deer populations on Vancouver Island associ-

ated with increasing densities of wolves, researchers did not document a decline in elk numbers during their study. However, researchers suggested that wolves might shift to heavier predation on elk if deer populations continued to decline and that this would impact elk populations.

The ecological role of large carnivores, and specifically wolves, in shaping forested landscapes through their influence on ungulates has been studied in various places around the world. European scientists studying wolves and hoofed mammals in the temperate forests of Poland and Belarus looked at several decades of data and determined that predation was maintaining herbivore populations below numbers that could be sustained just by habitat restrictions alone and that this dynamic contributed to a higher diversity and productivity in other parts of the ecosystem. In the Great Lakes region of North America, research from Isle Royal National Park documented increased growth rates in balsam firs (*Abies balsamea*) when released from heavy herbivory by moose when wolf populations increased and a reversal of this when wolf numbers dropped.

The removal of apex carnivores such as wolves has led to increases in ungulate densities and subsequent declines in plant, bird, and mammal biodiversity around the globe. In British Columbia, a study of islands where deer had been introduced, and where there were no wolves, showed a marked decrease in vegetation cover compared to areas with both deer and wolves.

With so many variables to consider, including various human hunting regimes, changes in other predator populations, and large-scale timber harvesting at low elevations on the Olympic Peninsula, picking out the specific

Riparian alder forest such as this one showing a lack of shrub and fern cover in the understory, as well as a lack of young alders, is common along much of the Hoh River.

effect of the disappearance of wolves on the current landscape may be impossible. One study, carried out on the west side of the peninsula, attempted to show a correlation between the cessation of recruitment of young cottonwoods and the period when wolves disappeared and ungulate population increased. This study relied on the coincidence of wolf extirpation with elk population increases and a subsequent failure of young cottonwoods to live beyond the sapling stage into mature canopy trees. However, because human hunting regimes and habitat changes were also going on during this same time, it is difficult to say if the absence of wolves alone is responsible for a lack of cottonwood sapling survival in parts of the peninsula.

The longevity of the dominant tree species in Pacific Northwest temperate rainforests creates an ecological structure that responds slowly to browsing pressure. And with mature old growth trees essentially immune from ungulate browsing pressure, the leverage that wolf predation can exert in such an environment is significantly less than in a grassland, steppe, or short-lived forest environment. Similarly, if high populations of elk are impeding the recruitment of western hemlocks in ancient forest in the Olympics, it could take several hundred years before this change would show up in the composition of the canopy trees in the forest.

However, in mountainous forest landscapes, elk and deer generally concentrate disproportionately on valley bottoms and riparian corridors. In the fall, migratory elk and deer drop to lower elevations to escape deep snowpack and join the resident animals that do not migrate. Along with having a milder climate than at higher elevations, riparian corridors in valley bottoms are particularly dynamic, as meandering and flooding rivers disturb the landscape, constantly creating primary succession opportunities. This combination of intensified population density and early successional-stage vegetation suggest a different picture of the ecological power of both ungulates and wolves on these critical habitats.

The return of wolves would likely lead to a reduction in total numbers of elk in some locations, and habitat use might shift. These changes could contribute to significant ecological shifts in specific parts of the Olympics. In places now heavily used by elk, cottonwood trees may regenerate in riparian forests, forest understory shrub density and diversity could increase, and in areas where elk activity has removed shrubs from along stream courses, bank erosion could decrease.

Wolves contribute to broad ecosystem changes most directly through their influence on the population size of their prey. Pretty straightforward—more elk, more browsing pressure; less elk, less browsing pressure. However a growing body of research suggests that wolves and other carnivores exert ecological influence in other, more subtle ways, through interactions with their prey. Even if the overall population of the prey species doesn't decline, prey species might change their behavior. This could include changes to the parts of the landscape they use as they balance predator avoidance with finding enough to eat. In the Rockies, researchers noted that following the reestablishment of wolves, in some instances elk made noticeable shifts in how they used the landscape, apparently in an attempt to reduce the threat of predation. Perhaps similar changes might occur with Olympic elk. However, in the Rockies, elk habitat use was least impacted during the winter when apparently the need for food and desire to go to the most profitable location to forage trumped any increased risk from predation by wolves there. If the same were true in the Olympics, then any behavioral shift in elk might not have much impact when elk are most congregated: during the winter in lower-elevation, south-facing slopes and, in early spring, when they shift to valley bottoms and riparian communities. Whether behavior shifts in prey species have contributed to many of the ecosystem changes documented in the Rockies is a subject of ongoing scientific research and debate.

The presence of a predator can also shift the behavior of a prey species in the balance of time the prey spends feeding in comparison to scanning for danger, or "vigilance." Theoretically, if an elk believes there is more imminent danger it will spend more time attempting to detect that danger so as to avoid it, at the expense of time foraging. With more and varied predators on the landscape, prey animals have to expend more energy attempting to avoid them, which will limit their ability to forage as efficiently. This could deter visits to high-risk areas or might reduce their overall health and lead to greater vulnerability to predation or starvation during the winter. A study comparing the vigilance of Rocky Mountain elk in Yellowstone, where wolves and grizzly bears are present and predation risk is high, and in Rocky Mountain National Park, where these two predators are absent and predation risk is significantly lower, did show a correspondingly higher level of vigilance in Yellowstone elk.

The human couch potato is a perfect example of this natural biological urge—the desire to situate ourselves in a safe and comfortable place where we can dull our awareness of the world around us and consume high-calorie foods with minimal effort. Humans have actually worked very hard to create such safe, predator-free environments for ourselves along with access to unlimited amounts of sugary and salty high-calorie foods. In some ways, human efforts to rid our surroundings of wolves might actually be traced back to this primal yearning. Several days after I visited the exclosures in the South Fork of the Hoh River, while camping up the main stem of the Hoh in the national park, I encountered the closest thing to couch potato elk behavior that I can imagine. My own behavior in the situation didn't rank much beyond this level either.

The main stem of the Hoh receives exponentially more visitors each year than the South Fork. A visitor center and huge campground provide car-accessible opportunities for many, while the Hoh River trail is the jumping off point for several popular backpacking routes. However, it was early in the spring, and the campground was almost empty. I pulled into a site overlooking the river. This part of the campground is built into a riparian alder forest. Like these forests for miles along the Hoh, the understory is dominated by grass with only patches of sword ferns and occasional deformed salmonberry bushes—signs of high elk abundance along the river.

I had scheduled to meet Marcus Reynerson that afternoon to scout for a wildlife-tracking class we would be teaching at that location. With some time to kill before he arrived, I sat in the relative comfort of the cab of my truck, warm and dry, downloading photos onto my laptop computer and relaxing. About an hour later I looked up and noticed Marcus and our two teaching assistants, Mark Kang O'Higgins and Lindsay Huettman, pulling into the campground, then taking a nearby parking spot. As they got out of the car, though, they weren't looking at me

A herd of Roosevelt elk grazing adjacent to the main stem of the Hoh River demonstrates exceptionally low vigilance with all its members keeping their heads to the ground for long periods of time. This allows them to be most efficient in feeding and also represents their very low sense of predation risk.

but rather beyond me. Marcus said something like, "Got some friends visiting?" I poked my head out the window. While I had been blissfully engrossed in my photographs, behind my truck a herd of about twenty elk had wandered into the campground and were grazing intently, a number of them within ten yards of my truck. Marcus, Lindsay, and Mark all approached my truck, and I got out to greet them. None of the elk looked up—they just kept their heads down and went right on grazing. We sat on the tailgate of my truck as they wandered through the campground feeding for the next hour.

Clearly, these elk did not perceive any threat from human presence. In actuality, the presence of people might be an attractant to these elk. In Banff National Park, in Alberta, researchers documented variations in the foraging patterns of elk related to human and wolf presence. In parts of a valley with limited human activity, elk activity was low and wolf activity was high. In these areas, willow growth, aspen recruitment, beaver density, and songbird abundance and diversity were all greater, while elk survival rate was much lower. In areas with high human presence, wolf activity diminished while elk survival rates were higher. The impacts of elk on willows and aspens were much higher. With a reduced food supply, beaver abundance was decreased. and with more limited refuge structure, songbird abundance and diversity was decreased. The elk appear to be using humans as a shield from the wolves, and the results cascaded through the landscape from elk to the plants they consumed (or didn't consume) to the other animals that also used those plants.

While there are no wolves in the Hoh, there are black bears and cougars. A couple of days later we found the tracks of a cougar a few miles upstream, on a much less visited section of the river. Farther up the Hoh, the elk acted wary and kept their distance from us. But even miles from the trailhead and the epicenter of human activity, we found the stream banks bare and the willows on the sandbars browsed into mangled clumps. Would the return of wolves to the large carnivore guild here bring about changes like those in other parks where wolves have returned? Perhaps time will tell.

THE MARMOT, THE COYOTE, AND THE WOLF

After years of tramping around the high mountains of the Pacific Northwest, I was used to the high-pitched whistle of a marmot but not at such close range. The noise-maker, a young Olympic marmot (*Marmota olympus*), was making its displeasure known to anyone who cared to hear. Its predicament was this: it had just been trapped and sedated, and a radio transmitter had been surgically implanted in its belly. As I accompanied a researcher carrying the marmot back to its burrow system to be released, I asked why marmots don't get a radio collar instead, like many other creatures. The answer: the size of their neck changes so drastically throughout the course of a year that it would be impossible to size correctly. Upon returning to the makeshift operating room, a tent strategically located in the shade of a subalpine tree island, wildlife veterinary surgeon Malcolm McCadie, clearly quite skilled in his trade, was working on his next client. At the helm of the project was Sue Griffin, who has been studying the plight of Olympic marmots for many years, first as her Ph.D. research and now out of a deep concern for the adorable and threatened species. I spent a day with Griffin and her crew while they were working at Obstruction Point in Olympic National Park.

Olympic marmots inhabit some of the harshest landscapes on the Olympic Peninsula, eking out their existence in a land covered by colossal amounts of snow for months out of the year, buffeted by the full brunt of storms rolling off the Pacific. To survive in this environment, these large ground squirrels take full advantage of the one season that is generally quite pleasant. During the brief summer months marmots reproduce, rear their young, and put on enough weight to survive six to eight months of hibernation. To accomplish this, marmots have a very particular social structure. A dominant male and one or occasionally two adult females form the nucleus of a colony. A colony includes several burrows, some with multiple entrances, and conspicuous throw mounds of soil. Females typically don't breed until four and a half years of age, and on average only about a third of females breed during a particular summer. Inconsistent breeding in females is probably due to the time it takes females to recover from the breeding effort. While over a third of young Olympic marmots never leave their natal home range, those who disperse do so generally after their third winter, typically driven off by their parents and subsequently establishing their own breeding territory. Even when they do disperse, females rarely travel more than a few hundred yards. This social structure allows for successful nurturing of young but creates a very low reproduction rate for a rodent species of their size and low ability to colonize suitable habitat at a distance from existing populations, two limitations that make their population more sensitive to predation.

The Olympic marmot, endemic to the Olympic Mountains of Washington State, is in decline. Coyotes, who first arrived in the range of this marmot after the collapse of wolf populations on the peninsula, are the leading cause of mortality.

A story about marmots and marmot researchers might seem a bit misplaced in a book on wolves. At least I would have thought so until I asked Griffin what she had discovered to be the causes of the decline of the Olympic marmot, a species naturally endemic to the high country of the Olympic Mountains. Many

endemic species, including the closely related Vancouver Island marmot, are threatened by habitat destruction and fragmentation. Not the Olympic marmot, though: essentially all of its habitat is protected within national park or federally designated wilderness areas adjacent to the park. So what was killing marmots? Overzealous park visitors chasing them around with cameras (they are a charismatic and conspicuous species in some popular parts of the park)? A novel disease introduced from elsewhere in the world? Climate change? Nope. Coyotes.

In an intensive study of Olympic marmots, Griffin discovered that predation was the cause of all mortalities that could be definitively accounted for. Of these, coyotes were responsible for more marmot deaths than any other predator. Based on recovered deceased radio-tagged marmots, carnivore scat analysis, and anecdotal observations of coyote hunting behavior around marmots, Griffin estimated that coyotes could be responsible for as much as 88 percent of mortality for Olym-

pic marmots and noted that she typically saw coyotes in the field while visiting marmot colonies. In the spring, when there is snow on the ground, she and her field crew would often find coyote tracks traveling across the landscape on a circuit going from one burrow system to the next.

The logical question, of course, is: why are Olympic marmots now threatened by coyotes? Have the coyotes become better marmot hunters? The answer is that, for the Olympic marmot, the coyote is a novel predator, having first arrived in the range of these marmots less than a century ago. Despite having literally all the habitat protection a threatened species could hope for, the Olympic marmot's prospects do not appear cheery. Griffin's research demonstrated that marmots, already a small population, were definitively in decline across their limited and shrinking range. According to her studies, the entire population is likely only about one thousand animals and shrinking annually.

RISE OF THE MESOCARNIVORE

One afternoon while walking back to our vehicles after a long day of fieldwork in northwest Montana, Cristina Eisenberg and I noticed the tracks of a coyote running along the edge of the hard-packed gravel road. In most places all that stood out were the deeply embedded nail marks in the roadbed. This, and the pattern of tracks—groups of four tracks stretched out in a straight line heading down the road—indicated that this animal was running fast. A coyote would most likely run flat out like this for one of two reasons: attempting to run down prey or fleeing for its life. Which was it?

We followed the tracks about twenty yards down the road where we discovered the answer in the form of another set of tracks, these of a wolf, also running hard. The wolf's trail cut onto the road and headed straight down it alongside the coyote's. The coyote was running for its life. Eventually the tracks disappeared off the road into substrate where we could not follow. What became of this particular coyote remains a mystery.

A coyote looks out from the brush on the edge of a wetland in the Puget Sound area.

When European Americans began settling in western Washington, coyotes were either absent or very rare around Puget Sound and absent entirely from the Olympic Peninsula until the first couple decades of the 1900s. The first documented reports of coyotes in western Washington are from the early 1900s. By the mid 1920s, their numbers were increasing rapidly on the Olympic Peninsula. Coyotes are still absent from most of the British Columbia coast and all of its islands including Vancouver Island. One report from 1929 declared coyotes were "invading western Washington as lumbering and clearing proceed." Trappers of the time noted that coyotes did not come into western Washington until wolves became scarce. Beyond western Washington, the range of coyotes expanded dramatically in North America following colonization, probably because of habitat alterations and destruction of wolf populations.

ABOVE: Coyotes, such as this one consuming a roadkilled deer, are not adept at killing adult deer or elk but will seek out such carcasses whenever possible, making coyotes competitors with wolves where the two species overlap.

OPPOSITE: Two coyotes travel through a Puget Sound wetland which is dominated by invasive reed-canary grass and Himalayan blackberry brambles. Like these plants, coyotes are a relatively recent addition to the Puget Sound lowlands.

The close lineage of wolves and coyotes produces the potential for a high degree of ecological overlap between the two species. Both are highly intelligent, opportunistic, and adaptable to a wide variety of ecological conditions. The basic social unit for both species is a breeding pair who generally stay together throughout the year.

The smaller coyote is highly omnivorous and much less effective at killing adult hoofed mammals. Wolves are hypercarnivorous, rarely eating anything but animal tissue. The social structure of coyote populations is adaptable and linked in part to their primary food source—under most conditions, small mammals, insects, and plant material, a diet that lends itself to smaller group sizes, with sub-adult animals dispersing when a new litter is born. With wolves, larger packs are common, accommodated by their reliance and skill in killing large prey that can feed more animals. Conversely, solitary wolves, or groups of wolves in the absence of large game, are able to survive on small mammals, foraging, and scavenging. In the absence of wolves but with access to big game carcasses, coyotes will form larger packs to most effectively make use of large food items.

Following land-clearing activities and the extirpation of wolves in the nineteenth and early twentieth centuries, coyotes expanded their range north and into coastal locations where they appear to have been absent before European-American settlement of the Pacific Northwest.

Just as wolves are members of the large carnivore guild, coyotes are members of a guild as well, the mesocarnivores. Mesocarnivores are small to medium-size carnivorous animals (up to about thirty-five pounds) including, in the Pacific Northwest, a numerous and diverse collection ranging from aquatic specialists such as river otters (*Lutra canadensis*) to generalists like the raccoon (*Procyon lotor*). Mesocarnivores are more likely to have an omnivorous diet than large carnivores are. Like their larger compatriots mesocarnivores play a critical, though very different, role in ecosystems. They can profoundly influence the population sizes of small mammals and birds on which they prey, and they can be important seed dispersers of the berries and fruit on which they forage.

Humans tend to be more tolerant of mesocarnivores, perhaps because we perceive them to be less threatening to our safety or food supplies. This tolerance, along with their relatively adaptable behavioral repertoire and diet, has allowed many species, including coyotes, to flourish in close proximity to humans. However, other mesocarnivores, such as the American marten (*Martes americana*) and Pacific fisher (*M. pennanti*), have been trapped extensively for their fur, and their populations have been decimated. When they scavenged poisoned carcasses left for wolves, cougars, and bears, mesocarnivores also suffered from the indiscriminate poisoning campaigns designed to kill these larger predators. Martens and fishers have not yet returned to large portions of their range in the region.

Mesocarnivores' relationship with larger carnivores, and specifically wolves, is complex. These smaller carnivores are often subject to predation and competition

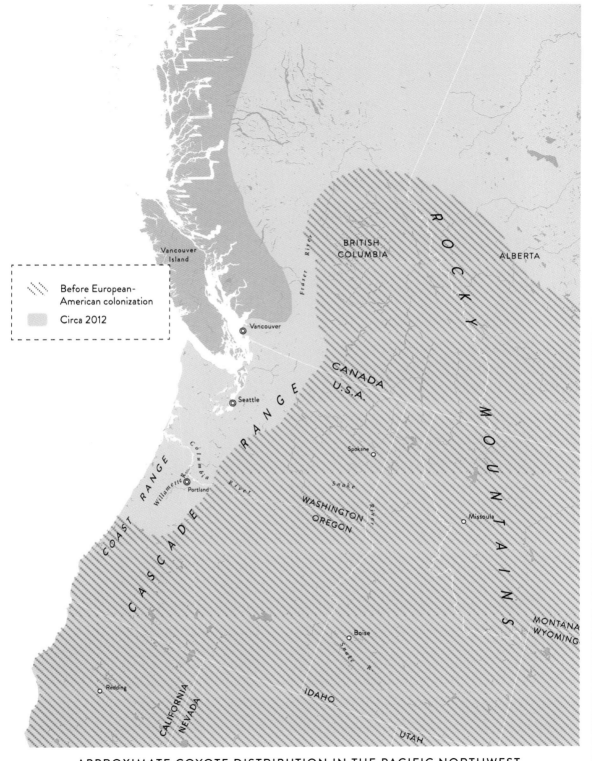

APPROXIMATE COYOTE DISTRIBUTION IN THE PACIFIC NORTHWEST

from larger carnivores as well as from other members of their own guild. Wolves opportunistically kill coyotes without necessarily eating them. Researchers in the North Cascades recently retrieved the remains of a radio-collared wolverine that had apparently been killed by a mountain lion close to a deer carcass.

Yet mesocarnivores often benefit from the presence of wolves. Some species are effective scavengers and make use of large carcasses produced by wolves and other large carnivores. Other species, such as red foxes (*Vulpes vulpes*), benefit indirectly from wolves. In some areas coyotes depress fox populations (for the same reasons that wolves depress coyote populations). When wolves enter these landscapes and coyote populations are reduced, fox populations may rebound. With the decrease in coyotes, other mesocarnivores may also have greater access to shared food sources.

Mesocarnivore and scavenger consumption of carcasses killed by wolves can profoundly affect the rate of kills that wolves must make to sustain themselves. In this way, mesocarnivores can both increase nutrient flows within a community and increase the ecological influence of wolves. Though their role as a provider of meat may be a service to the overall productivity of the landscape, this is not a role that wolves take on out of a good-natured love of all their fellow carnivores. To the contrary, wolves defend their kills aggressively. In the Rocky Mountains, coyotes appear to benefit more than any other mesocarnivore from wolf-killed carcasses of hoofed mammals. Not surprisingly, they also get the most negative attention from wolves as well.

Wolves are involved with two types of competition with other carnivores: exploitive and interference. Exploitive competition involves two animals competing for the same resources, for instance, wolves, coyotes, and ravens all competing for the remains of a winter-killed deer. Interference competition involves one animal or species attempting to deter another from accessing a particular resource, often through aggression. For example, wolves from one pack may kill a transient wolf in their territory, or wolves may harass or kill coyotes attempting to scavenge on a carcass the wolves killed or would like to claim. Typical of strong interference competition, wolf and coyote population density is often inversely related. Wherever the two species overlap, it is relatively common for wolves to kill coyotes, often not consuming them. In the Rockies, coyotes are most often killed by wolves in winter on wolf-killed ungulate carcasses.

Because of this dynamic, coyotes have likely benefited more than any other mesocarnivore from the reductions of wolf populations regionally. Released from competition with wolves, coyotes were able to expand their range and population size in western Washington. Land-clearing activities for agriculture and logging created more open landscapes with abundant small mammal populations that are typically a primary food source for coyotes. Theoretically, resurgent wolf populations regionally could initiate a reversal of the coyote's fortunes. It is unlikely that wolves would eliminate coyotes entirely from areas they have expanded into as the advance of coyotes coincided with both wolf removal and large habitat alterations. These habitat alterations, in many instances, have created totally new landscapes where coyotes have learned to thrive. West of the Cascades, forests converted to agricultural land, clearcuts, and early succession forests produce higher abundances of small mammals, insects, and plant foods that coyotes focus on. Coyotes have successfully colonized urban areas in the Northwest, now living in the most urban parts of Seattle. Even in a city as progressive as Seattle, it is unlikely that humans would tolerate wolves in such close proximity. Reductions of the overall population of coyotes, and possibly extirpation from specific habitat types, is more plausible.

What sort of impacts this would have on coyotes who frequent Olympic marmot habitat is not known. Wolves also hunt marmots from time to time. However, wolves, specialists in hunting large game, consume fewer small prey species than coyotes do. In addition, Olympic marmots and wolves coexisted for thousands of years in the Olympics, and marmot habitat has remained basically unchanged. In sum, this would suggest that a resurgent wolf population, which would depress coyote populations, would likely be a net gain for the marmot.

While the direct impacts of wolves and other apex carnivores on a landscape can be very apparent, their subtle influences on a region's ecology may not be obvious for decades after their removal. In the story playing out between coyotes and Olympic marmots we see a pattern that has accompanied the widespread removal of wolves and other large carnivores around the globe.

ANOTHER MARMOT PARADOX: WOLVES AND THE PLIGHT OF THE VANCOUVER ISLAND MARMOT

Just north of the Olympic peninsula, across the Strait of Juan De Fuca, Vancouver Island is home to the Vancouver Island marmot (*Marmota vancouverensis*). Closely related to the Olympic marmot and considered by many to be Canada's most endangered mammal, these marmots number three to four hundred animals, about one hundred of them in a captive breeding program. There are no coyotes on Vancouver Island and never have been, but predation is the chief source of mortality for the Vancouver Island marmot, as it is for the Olympic marmot. Who's the culprit here? In one study, wolves accounted for 38 percent of mortality, followed by mountain lions, which accounted for another 21 percent. Why are wolves and cougars, two predators that the Vancouver Island marmot has lived with for thousands of years, now possibly a significant factor in their flirtation with extinction?

It's the familiar story: survival by avoidance. Vancouver Island marmots, like the Olympic marmot, live in subalpine and alpine meadows at high elevation. All was well until humanity's lust for cellulose came clamoring into their quiet mountain kingdom. Between the 1940s and 1980s road building and logging operations transformed the forests adjacent to the marmots' meadow habitat. Removal of the ancient forests didn't itself send the marmots into decline, however, as some marmots dispersed into clearcuts and established colonies in this new, open habitat.

For several decades after being cut, clearcuts can sustain much higher densities of deer and other hoofed mammals. On Vancouver Island, in locations that have not been subject to timber harvest, wolf activity tends to concentrate in valley bottom and lower-elevation habitats. But in high-elevation clearcuts, as deer populations increased rapidly, wolves and cougars followed. As with caribou in the interior, marmots became a secondary and opportunistically taken prey for these two large carnivores. More recently golden eagle predation on marmots has increased, possibly as this predator increased in numbers following the introduction of feral domesticated rabbits and eastern cottontails in some locations. Furthermore, while marmots have been documented to set up residence in clearcuts, survival in these environments was lower than in natural meadow habitat. Clearcut habitat may be acting as a population sink for Vancouver Island marmots and preventing the recolonization of higher-quality but more distant natural meadow habitat.

In another bizarre twist of fate, provincial game managers, desiring to increase deer populations for human hunting, initiated a controversial predator control program on the island during the 1980s. In the short term these efforts were successful, not just in increasing deer populations, but in allowing marmots to increase in numbers and distribution as well. Of course, once predator control ceased, the increase of deer likely only exacerbated the issue for marmots and their populations began dropping again.

As with mountain caribou, we are left with a dilemma: with the habitat changes we have wrought, deer, wolf, and mountain lion populations will take at least decades to restabilize naturally at numbers low enough to give marmots relief from increased predation. This is assuming that no further timber harvesting occurs in their habitat, though only about a third of the area is currently protected in ecological reserves. In the meantime, their population would have almost certainly slipped into extinction if it weren't for the ongoing captive breeding program.

GETTING BACK TO
THE OLYMPICS

A 2004 study looking at the feasibility of reestablishing wolves on the Olympic Peninsula predicted that a wolf population could be sustained there. Probably overly conservative in its estimates of suitable habitat for wolves, the study excluded many portions of the peninsula outside of the national park based on assumptions about wolves' inability to use landscapes with specific road densities. The accuracy of this model, based on wolves in an expanding population in the Great Lakes, was subsequently challenged. It appears that once this Great Lakes wolf population began saturating higher-quality habitat, adjacent, more heavily roaded landscapes began to be occupied as well. Further, road data often does not discriminate between open and gated roads which don't provide motor vehicle access to the public. This is an important distinction in the Olympics where lower-elevation forests outside of the park tend to be private timberland with highly restricted public access.

Of course key to the feasibility of a return to the Olympics would be wolves actually being able to make it back there, a feat virtually at the whim of humans for the foreseeable future. Past feasibility studies were based on the idea that humans would translocate wild wolves from elsewhere to the Olympics. In such a scenario enough animals to start a viable breeding population would be strategically released in ideal locations, likely within the park. From these original introduced animals a growing population would expand into adjacent areas and create a self-sustaining population. Washington's wolf recovery plan allows for translocation of wolves from one part of the state to another (but not from places outside of the state into it). Theoretically, under this management plan, wolves could be moved into the Olympics. However, attempts to do this would likely be met with stiff resistance from some segments of the human population in the Olympics.

As for wolves returning naturally to the Olympics, this will be a long haul even for the wide-ranging wolf. The most likely route would be from the southern Washington Cascades to the Olympics through southwestern Washington. Between the quality habitat in the Cascades and Olympics is a large section of landscape that would not likely support a stable population of wolves. This break between the southern Washington Cascades and the Olympics was recognized as the largest obstacle for dispersal of carnivores anywhere in the state, though not out of reach for an animal with the dispersal capacity of a wolf.

EPILOGUE

Brave New World

THE SKYKOMISH RIVER, on the west slope of the Cascades northeast of Seattle, flows out of the mountains and eventually into the Snohomish River, which itself shortly drains into Puget Sound. One gray and wet fall day, while teaching a wildlife tracking class on the river just east of the town of Sultan, my students and I made an unexpected discovery.

At our location the river had begun its transition from a steep wild mountain stream to a meandering, slower-moving river wending its way through a valley converted from forest to agriculture well over a century ago. The valley bottom is flat here and several miles across, with the steep foothills of the Cascades rising up on either side. The close mountainsides are scarred with logging roads, large blocks of clearcuts, and the deep vertical skidder tracks where logs were dragged up to landings on the roads. Still, with a backdrop of snow-covered Cascade peaks, the cold clear water, and the smell of rotting fish along the riverbanks, a sense of wild grandeur pervades the landscape.

When the salmon are spawning on the river, the wildlife tracking is definitely more exciting. On this day, my students and I picked up the tracks of a pair of coyotes shortly after entering the floodplain forest. We followed the tracks out toward the river. The cottonwoods and alder gave way to sand and river cobbles with the river beyond. A bald eagle, perched on the other shore, pecked at the carcass of a fish larger than itself, as several crows looked on from a respectful distance. The smell of rotting fish, evident on many of the region's rivers at this time of year, was strong.

The river had been dropping in volume, stranding a dozen or more chum salmon on the banks. Coyote tracks, laid down in the freshly deposited river silt, continued the story. They led from one carcass to the next. A number of the carcasses were headless. I had never heard of coyotes feeding on salmon in this way, though it is identical to how wolves farther north on the coast feed on them.

There were likely no coyotes along this river prior to European-American settlement of the area. My students and I spent the day on the banks of the river, imagining the changing landscape, the glaciers millennia ago that scoured the land, carving the mountainsides, the subsequent dense forests of western red-cedar, Sitka spruce, and western hemlock that dominated the landscape, and the salmon that returned to the newly ice-free rivers. Then wolves and grizzly bears likely prowled this riverbank hunting deer and Roosevelt elk and feasting on the abundant salmon runs, taking advantage of receding stream flows trapping fish in sloughs and side channels in the wild, braided river channel. These carnivores were joined by the Snoqualmie people, who also made their living pulling fish from the river and hunting in the valley's forests. With the arrival of European Americans, the big trees retreated into the inaccessible mountains, replaced by agricultural fields and livestock. The wolves and grizzly bears were hunted, trapped, and poisoned to defend cattle and sheep, the elk market-hunted to extinction locally. Following the disappearance of their cousin and nemesis, the coyotes emerged. Under the 1855 Treaty of Port Elliot, the Snoqualmie Tribe, whose office is now an hour's drive down the highway from this section of the river, lost control of all of their traditional lands, though they retained fishing rights. As is the case with most residents of the Puget Sound area today, most of my students and I arrived in the region from someplace else only recently. At the culmination of all these events, we explored here, piecing together the complex story written into the landscape, ending with a collection of headless fish on the banks of the river. What stories will unfold in the next chapter of this place? Like the river, our beliefs and behavior as a species are in constant motion, and with this movement we transform the world around us as well as our understanding of it.

PREVIOUS: A chum salmon which has been fed on by coyotes in a manner similar to that of wolves. Skykomish River, Washington.

NOTES

INTRODUCTION

Page
15. their peak census: IDFG and Nez
    Perce Tribe 2012.
15. about 100 wolves: IDFG 2009
15. about thirty: WDFW 2012a.
15. as of December 2011: Morgan 2012.
15. in the state: ODFW 2010.
15. state endangered species list: Wiles,
    Allen, and Hayes 2011.
15. 375 wolves: IDFG 2012.
15. the entire state: IDFG and Nez Perce
    Tribe 2012.
20. chemicals in their tissues: McAr-
    thur et al. 1993; Gonzalez-Hernandez
    2000.
21. shifts in the planet's climate: Meehl
    et al. 2007.
21. every ecosystem: Terborgh and Estes
    2010; Hoekstra et al. 2005.
21. percentages of the earth's existent
    species: Thomas et al. 2004.
21. a sixth great extinction event:
    Leakey and Lewin 1995.
22. resilience in the face of disturbances:
    Terborgh and Estes 2010.

CHAPTER 1

34. recaptured and taken out of the wild:
    Almack and Fitkin 1998.
35. about 355 square miles: WDFW,
    unpublished data.
35. nine or ten animals, and other
    observations of the Lookout pack:
    Ray Robertson (wildlife contractor,
    Okanogan-Wenatchee National For-
    est, Carlton, Washington), interviews
    and email correspondence, April
    2010 to October 2011.
35. habitat throughout the year: Carroll
    et al. 2001.
35. eating a hoary marmot: Ray Robert-
    son, interviews and email correspon-
    dence, April 2010 to October 2011.

36. "seasonal food source": John Rohrer,
    email correspondence, September
    2011.
36. decaying fish: Wipfli, Hudson, and
    Caouette 2003.
37. should be protected in the state:
    WDFW 2011.
38. livestock business: Kellert 1996.
39. outside of preserved areas: Forman
    2004.
44. the lake in 1990: Almack and Fitkin
    1998.
44. North Cascades and beyond: Almack
    and Fitkin 1998; Two gray wolf packs
    1990; Fritts 1992; Scott Fitkin (wild-
    life biologist, WDFW, Winthrop,
    Washington), interview, 17 June 2010.
48. wolves in the region: Okanogan
    County District Court 2009.
48. illegal in the state: Campbell 2010.
48. depredation was documented:
    WDFW 2012b.

CHAPTER 2

54. the rest of the time: Gittleman 1989.
54. common ancestor of both: Wang and
    Tedford 2008.
54. social hunter of large game: Bekoff
    1978.
54. history of the canid family: Wang
    and Tedford 2008.
55. half of its large mammal species:
    Lange 2002.
55. collected from Oregon: Nowak 2003.
55. prey that remained: Nowak 2003.
55. more prone to fit the model than
    smaller ones: Meiri and Dayan 2003.
55. in the Pacific Northwest: Hatler,
    Nagorsen, and Beal 2008; Jim
    Holyan (wildlife biologist, Nez Perce
    Wildlife Program, Lapwai, Idaho),
    email correspondence, May 2011.
55. five and a half feet: Hatler, Nagorsen,
    and Beal 2008.
56. Idaho in the past: Cockle 2011.

57. more rounded ears than coyotes: Mech 1974.

57. more arid environments: Bekoff 1978.

60. and related elements: Elbroch 2006; Wang and Tedford 2008.

60. relative to their size: Peterson and Ciucci 2003.

60. better than ours: Asa and Mech 1992.

62. that of humans: Asa and Mech 1992.

63. engage with prey species: Wang and Tedford 2008.

63. elk and deer: Peterson and Ciucci 2003.

64. stride of the animal: Howell 1944.

64. excess food from kills: Wang and Tedford 2008.

65. typical for mountain lions: Wang and Tedford 2008; Peterson and Ciucci 2003.

65. social hunters: Wang and Tedford 2008.

70. environmental conditions: Peterson and Ciucci 2003.

70. once food is available: Peterson and Ciucci 2003.

70. den or rendezvous site: Peterson and Ciucci 2003.

70. separately or together: Kreeger 2003.

70. reach four years: Fuller, Mech, and Cochrane 2003.

71. second half of winter: Kreeger 2003.

72. about sixty-three days: Mech 1974.

72. through the summer months: Ian McAllister (founder and director, Pacific Wild, Denny Island, British Columbia), telephone interview, 11 June 2010.

72. carnivores their size: Oftedal and Gittleman 1989.

72. stable or saturated populations: Fuller, Mech, and Cochrane 2003.

72. higher survival rate: Fuller, Mech, and Cochrane 2003.

72. typically, three years: Aune, Mace, and Carney 1994.

72. age of sixteen: Mech 1974.

73. mortality for adult wolves: Fuller, Mech, and Cochrane 2003; Musiani, Boitani, and Paquet 2009.

73. mortality for wolves: Wydeven et al. 2009.

73. is another wolf: Peterson and Ciucci 2003.

73. and vice versa: Fuller, Mech, and Cochrane 2003.

73. most at risk: Peterson and Ciucci 2003.

73. across the continent: Williams and Barker 2001.

73. wolves in that region: Smith, Peterson, and Houston 2003; Hamashige 2006.

74. breeding territory and pack: Fuller, Mech, and Cochrane 2003.

74. control of the territory: Mech and Boitani 2003.

74. researchers documented: McAllister 2007.

75. up to eleven: Fuller, Mech, and Cochrane 2003.

75. primary prey species: Mech and Boitani 2003.

75. moose or deer: Fuller, Mech, and Cochrane 2003.

76. pack size in general: Fuller, Mech, and Cochrane 2003.

76. adjacent pack's territory: Mech and Boitani 2003.

77. into a smaller space: Mech and Boitani 2003.

77. Vancouver Island: Scott and Shackleton 1982.

77. in northeast Oregon: ODFW, unpublished data.

77. 190 square miles: Kent Laudon (wolf management specialist, Montana Fish, Wildlife and Parks, Helena, Montana), email correspondence, 2011.

77. 225 square miles: Jim Holyan, email correspondence, May 2011.

77. about 112 square miles: Chris Darimont (research scientist, Raincoast Conservation Foundation, Sydney, British Columbia), telephone interview, 22 July 2010.

77. packs in the North Cascades: Scott Fitkin, interview, 17 June 2010.

78. and wolves: Darimont and Paquet 2000, 2002; McAllister 2007.

78. large prey: Carbone et al. 1999.

CHAPTER 3

86. half of all such rainforests: Wolf, Mitchell, and Schoonmaker 1995.

87. in wolf pelts: WDFW 2011.

87. British Columbia and Alberta: Hansen 1986.

90. public buildings in the state: Charlie Raines (checkerboard lands historian), telephone interview, 30 October 2011.

91. since the 1980s: Toweill and Vecellio 2004; Base, Zender, and Martorello 2006.

92. leg-length-to-body-size ratio: Janis 2002.

92. prey for wolves: Seip 1992.

92. Diamond pack's range: Stotyn, McLellan, and Serrouya 2007; Seip 1992; Bergerud and Elliot 1986.

92. boreal forests of Ontario: Bowman et al. 2010.

93. meadows for the summer: Apps et al. 2001; Stotyn, McLellan, and Serrouya 2007.

94. density and therefore detectability: Bergerud and Page 1987, Smith et al. 2000.

94. vulnerable to predation: Smith et al. 2000.

94. attention of wolves: Stotyn, McLellan, and Serrouya 2007.

94. documented in our region: Kinley and Apps 2001.

94. shrinking caribou populations: Stotyn, McLellan, and Serrouya 2007; Seip 1992; Wittmer et al. 2005.

95. in their habitat: British Columbia Ministry of Agriculture and Lands 2007.

95. caribou conservation efforts: Chris Ritchie (Fish and Wildlife Recovery Implementation Manager, Prince George, British Columbia), telephone interview, 1 November 2011.

95. protect and restore caribou habitat: Orlando 2009.

95. climax forest tree species: Michael Borysewicz (wildlife biologist, Colville National Forest, Sullivan Lake, Washington), interview and email correspondence, November 2011.

98. reproductive rates for grizzly bears: Ballard, Carbyn, and Smith 2003.

98. killed the deer: Ray Robertson, interviews and email correspondence, April 2010 to October 2011.

99. kill rate of these large cats: Ballard, Carbyn, and Smith 2003.

99. risks of being killed: Berger 2010.

99. impacts on prey: Griffen et al. 2011.

102. typical in the past: Michael Borysewicz, interview and email correspondence, November 2011.

CHAPTER 4

113. role in ecosystems: Marquis 2010, Berger 2010.

114. the right conditions: Peterson and Ciucci 2003.

114. fewer wolves: Gittleman 1989.

114. separated from the group: Peterson and Ciucci 2003.

114. couple hundred yards: Mech 1970.

115. at the hind quarters: Stahler, Smith, and Guernsey 2006.

115. asphyxiation: Halfpenny 2003.

115. dug for this purpose: Peterson and Ciucci 2003.

115. as possible: Peterson and Ciucci 2003.

115. multiple wolves to attack: Peterson and Ciucci 2003.

116. of African lions: Gittleman 1989.

116. wolf hunting methods: Mech and Peterson 2003.

116. deer-sized animal: Fuller 1989.

116. muscles of the hind legs: Stahler, Smith, and Guernsey 2006.

116. bones and hide remain: Peterson and Ciucci 2003.

117. according to estimates: Stahler 2000.

117. half of a moose carcass to scavengers: C. Promberger, unpublished data, cited in Peterson and Ciucci 2003.

117. their social nature: Vucetich, Peterson, and Waite 2004.

117. WDFW estimated: WDFW 2011.

117. wolves' summer diet: Stahler, Smith, and Guernsey 2006.

120. to find and subdue prey: Mech and Peterson 2003, Stahler, Smith, and Guernsey 2006; Smith et al. 2000.

121. vulnerability of prey: Mech and Peterson 2003.

121. biological and behavioral strategies: Mech and Peterson 2003.

121. detect approaching wolves: Hebblewhite, Merrill, and McDonald 2005.

121. likelihood of detection: Creel et al. 2005; Muhly et al. 2010.

121. in dense cover: Kunkel and Pletscher 2001.

123. winter pass: Mech and Peterson 2003; Mech 2007.

123. higher deer abundance: Kunkel et al. 2004.

123. included domestic stock: Fritts et al. 2003.

124. 50 to 150 years or more: Berger, Swenson, and Persson 2001.

124. as a predation threat: Berger 2010; Berger, Swenson, and Persson 2001; Kauffman et al. 2007.

124. where wolves are absent: Berger, Swenson, and Persson 2001.

124. refuges from predation: Kauffman et al. 2007.

124. "of the antelope": Jeffers 2002, 563.

124. cats and rats: Blackburn et al. 2004.

124. around the globe: Clavero and Garcia-Berthou 2005.

126. as low as 1 to 4 percent: Mech and Peterson 2003.

126. factor in a successful hunt: Mech and Peterson 2003.

126. easier time finding them: Hebert et al. 1982.

127. with Cristina Eisenberg: Eisenberg 2010.

127. initial encounter occurred: Eisenberg 2010.

127. open country species: Geist 1982.

127. conceal themselves: Geist 1981.

127. in the open: Kunkel and Pletscher 2001.

127. absent on islands: Darimont, Paquet, and Reimchen 2008.

128. of the wolves' diet: Hebert et al. 1982.

128. most common prey for wolves: Hebblewhite, Pletscher, and Paquet 2002.

128. eight species of ungulates: Stahler, Smith, and Guernsey 2006.

128. and 14 percent elk: Kunkel et al. 1999.

128. important non-ungulate food source for wolves: Peterson and Ciucci 2003.

128. parts of our region: Price et al. 2005; Hebert et al. 1982.

128. are most vulnerable: Peterson and Ciucci 2003.

129. across the region: Verts and Carraway 1998.

129. impounded by beaver dams: O'Neil et al. 2001.

129. raising water tables: Olson and Hubert 1994.

132. more important than elk: Kertson 2010.

132. in the summer: Hebert et al. 1982.

132. poor habitat for the species: Darimont, Paquet, and Reimchen 2009.

132. primarily wild berries: Carrera et al. 2008; Darimont et al. 2004.

133. den site miles away: Ray Robertson, interviews and email correspondence, April 2010 to October 2011.

CHAPTER 5

141. rainforests in the world: Wolf, Mitchell, and Schoonmaker 1995; Della-Sala 2010.

141. miles into the continent: Raincoast 2010.

141. plight of the region: Armstrong 2009.

144. sea wolves: McAllister and Read 2010.

146. around salmon-bearing streams: National Research Council 1996.

147. of their birth: Dittman and Quinn 1996.

147. glaciers melted and rivers returned: National Research Council 1996.

147. within a couple of weeks: National Research Council 1996.

151. primary item in the diet: Darimont, Paquet, and Reimchen 2009.

152. shore for consumption: Darimont, Reimchen, and Paquet 2003.

152. approach from any direction: Douglas Brown (field station manager, Raincoast Conservation Foundation, Denny Island, British Columbia), interviews and conversations, September 2011.

152. if untreated: Headley et al. 2011.

152. kidneys of the fish: Darimont, Reimchen, and Paquet 2003.

152. portion of the fish: Darimont, Reimchen, and Paquet 2003.

153. body as well: Reimchen 2000.

153. coastal wolves, however: Szepanski, Ben-David, and Van Ballenberghe 1999.

153. was in the past: National Research Council 1996.

153. and even squid: McAllister 2007;
Darimont et al. 2004.

153. additional 16 percent: Darimont,
Paquet, and Reimchen 2009.

154. 82 percent of wolves' diets: Darimont, Paquet, and Reimchen 2009;
Hebert et al. 1982.

154. 28 percent of wolves' diets: Hebert et
al. 1982.

154. wolves' diet increases: Darimont,
Paquet, and Reimchen 2008.

154. beavers in this case: Milne, Harestad,
and Atkinson 1989.

154. little forage for ungulates: Person et
al. 1996; Happe et al. 1990.

154. Alaska and British Columbia: Bob
Hansen (wildlife-human conflict
specialist, Pacific Rim National Park,
Ucluelet, British Columbia), interview, 12 July 2010.

154. salmon consumption was higher:
Kohira and Rexstad 1997.

154. destruction of spawning habitat:
National Research Council 1996.

154. coastal creeks and rivers: National
Research Council 1996.

156. kill ravens at carcasses: Ballard, Carbyn, and Smith 2003.

158. roaded and logged: Friends of Clayoquot Sound 2011a.

158. old growth forest stands: Friends of
Clayoquot Sound 2011b.

159. predator control activities: Muñoz-
Fuentes et al. 2009b.

159. high elevations, were used: Hebert et
al. 1982.

159. locations in North America: Scott
and Shackleton 1982.

159. captured and collared: Scott and
Shackleton 1982.

160. or along the coast: Scott and Shackleton 1982.

162. bears at such locations: McAllister
2007; Chris Darimont, telephone
interview, 22 July 2010.

162. larger animal dominating: Palomares and Caro 1999.

162. next closest piece of land: Chris
Darimont, telephone interview, 22
July 2010.

163. high-quality spawning habitat:
Krkosek et al. 2007.

163. on a similar scale: Raincoast Conservation Foundation 2010.

166. made use of them: McAllister 2007.

167. low points in them: Douglas Brown,
interviews and conversations, September 2011.

167. return to specific streams: Routledge
2001.

CHAPTER 6

172. glaciers in our region: Suttles and
Ames 1997; Lange 2002.

173. the dire wolf: Lange 2002.

173. Amur tigers: Miquelle et al. 2005.

173. in sub-Saharan Africa: Woodroffe
and Ginsberg 2005.

173. advances and retreats of continental
ice sheets: Nowak 2003.

174. plant and animal communities: Littell et al. 2009.

174. constant flux: Krebs 2009.

174. about 20 degrees latitude: Fuller,
Mech, and Cochrane 2003.

174. along the coast: Hatler, Nagorsen,
and Beal 2008.

174. in forested regions: Verts and Carraway 1998; Johnson and Cassidy
1997; Hatler, Nagorsen, and Beal
2008.

176. west of the Cascades: Verts and Carraway 1998; ODFW 2010; Csuti et al.
2001.

176. Columbia Plateau altogether: Dalquest 1948.

176. two academic reports: Verts and Carraway 1998; Csuti et al. 2001.

176. for their demise to be noted: ODFW
2010.

176. landscape-burning: Agee 1993.

176. plant-cultivation techniques: Deur
and Turner 2005.

177. over 300,000 years old: Verts and
Carraway 1998.

177. explorers and settlers: Schoen 1972.

177. Dall sheep, a subarctic species:
Schoen 1972.

177. in some locations: Kay 1994, but see
Yochim 2001.

177. abundance of ungulates: Agee 1993.

177. beaver and other terrestrial furbearers: Scott and De Lorme 1988.

177. nearly four decades: Laufer and Jenkins 1989; WDFW 2011.

178. removed from the USFWS endangered species list for the first time:
Mack et al. 2010.

178. to nothing by 1856: Hudson's Bay archives as cited in Laufer and Jenkins 1989.
178. west slope of the Cascades: ODFW 2010.
178. same time in Washington: WDFW 2011.
178. including Vancouver Island: WDFW 2011; ODFW 2010; Hatler, Nagorsen, and Beal 2008.
178. or elsewhere in North America: Muñoz-Fuentes et al. 2009a.
178. reoccupying Vancouver Island: Muñoz-Fuentes et al. 2009b.
178. into the Washington Cascades: WDFW, unpublished data.
180. 1995 and 1996: Fritts et al. 1995.
180. this reintroduction effort: ODFW 2011.
180. into central Idaho: Hampton 1997.
180. were inaccurate: Wayne 2010.
180. all of North America: Nowak 2003; Hatler, Nagorsen, and Beal 2008.
180. northern Rockies of the United States: Nowak 2003.
180. occurring naturally: Brewster and Fritts 1995.
182. absolute proximity and genetics: Wayne 2010.
182. similar conclusions: Muñoz-Fuentes 2009a.
183. Washington, Oregon, and California: Larsen and Ripple 2006; Carroll et al. 2001, 2006; Carroll 2007; Oakleaf et al. 2006; USFWS et al. 2008.
183. heavily subsidize their diets: Darimont, Paquet, and Reimchen 2008, 2009.
184. in the Rocky Mountains: Carroll et al. 2001.
184. of British Columbia: Stotyn, McLellan, and Serrouya 2007.
184. ability to access prey here: Carroll et al. 2001.
185. prey on livestock: Bangs et al. 2009.
185. distribution in the Pacific Northwest: Hatler, Nagorsen, and Beal 2008; WDFW 2011.
189. long distances before settling: Mech and Boitani 2003.
189. natal territory annually: Fuller, Mech, and Cochrane 2003.
194. persecution by humans: Musiani, Boitani, and Paquet 2009.

194. increase moose populations: Stotyn, McLellan, and Serrouya 2007.
194. populations have been expanding: Boitani and Ciucci 2009.
194. thousands of interested humans: Halfpenny 2003.
194. road ecology: Forman et al. 2003.
194. hunting, trapping, and poisoning: Fritts et al. 2003.
194. proximity to roads: Fuller, Mech, and Cochrane 2003.
194. similarly to roads in this regard: Darimont and Paquet 2000.
194. in the Northwest: Fritts et al. 2003, Bangs et al. 2009.
195. detecting game: Paquet et al. 2010.
195. avoid this disturbance: Darimont and Paquet 2000.
195. reestablishment in several areas: Musiani, Boitani, and Paquet 2009.
195. from establishing breeding populations: Bangs et al. 2009.
195. a very long time: Wiles, Allen, and Hayes 2011; ODFW 2010.

CHAPTER 7

201. extirpation in the 1940s: ODFW 2010.
203. support the recovery of wolves: ODFW 2010.
204. "left in this area": Crandall 2011.
204. "users off of the land": Hubbard 2011.
204. "maintain native biodiversity": Fascione 2006, 8.
205. adult elk in the area: ODFW 2003.
205. regulated elk hunting season: ODFW 2003.
207. on Blue Mountains forests: Riggs et al. 2000.
212. license and tag sales: Idaho Legislature 2011.
212. 1900s in North America: Mahoney 2004.
212. overall ecological integrity: Nelson et al. 2011; Dunlap 1990.
212. one year old in the state: Rachael 2010.
212. from open shrublands: Rachael 2010.
213. were more likely: Griffen et al. 2011.
213. mortality in this herd: IDFW 2010.
213. Idaho wildlife managers: Groen 2010.

215. "mostly unknown": Rachael 2010, 34.

215. salmon runs in the area: Monte Miller (range technician, Boise National Forest, Boise, Idaho), interview, 22 October 2011.

215. increases in moose populations: Berger et al. 2001.

215. return of wolves: Beschta 2003; Ripple and Beschta 2003; Ripple, Rooney, and Beschta 2010; Berger and Smith 2005.

215. populations to increase: Hebblewhite et al. 2005; Nietvelt 2001.

215. still being debated: Kauffmann, Brodie, and Jules 2010.

215. "health of ecosystems": Fascione 2006, 8.

216. "little exposure to large wild mammals": Al Thieme (Portland, Oregon), email correspondence, 24 September 2011.

218. tribes to the south: Josephy 2007.

218. to Puget Sound to market: Prater 1983.

218. for economic viability: Donahue 1999.

218. abnormally high population levels: Berger 2006; Donahue 1999.

218. 1.1 to 1.2 million cattle and 46,000 to 58,000 sheep: WDFW 2011.

218. 1.4 million cattle and 217,000 sheep: ODFW 2010.

218. grazed on public lands: ODFW 2010.

218. caused by overgrazing: Donahue 1999.

219. federal government for them: Donahue 1999.

219. part of the landscape: Kauffman et al. 2001.

220. against the USFS: U.S. District Court 2011.

221. agency's ecological goals: U.S. District Court 2011.

222. far from uniform: Bangs et al. 2009.

222. grazing in the Northwest: Donahue 1999.

224. millions of dollars annually: Wildlife Services 2011a.

224. 748 red-tailed hawks: Wildlife Services 2011b.

224. state's entire wolf population: Holyan et al. 2011.

224. "not simply to kill carnivores": Berger 2006, 751.

224. for livestock production: Bangs et al. 2009.

225. constructed at the ranch: ODFW 2009a.

225. prevent further depredations: ODFW 2009b.

225. wolf recovery efforts: Musiani, Boitani, and Paquet 2009.

225. from the presence of wolves: Wiles, Allen, and Hayes 2011; ODFW 2010.

225. these areas at all: Muhly et al. 2010.

226. social identity: Donahue 1999.

226. agricultural tax subsidies: Donahue 1999.

227. integrity of the region: Muhly and Musiani 2009.

227. forest cover instead: Muhly et al. 2010.

227. to active predation: Fritts et al. 2003.

228. these two states: WDFW 2011; ODFW 2010.

228. the study period: Muhly and Musiani 2009.

228. averages about 231,000: Galle, Collinge, and Engeman 2009.

228. killed by wolves: Holyan et al. 2011.

228. cattle than wolves do: WDFW 2011.

228. with greater brush cover: Bradley and Pletscher 2005.

228. at a greater distance: Oakleaf, Mack, and Murray 2003; Bradley and Pletscher 2005.

228. use of guard dogs: Stone et al. 2008.

228. defensive strategies: Mech and Peterson 2003.

229. sometimes months later: Mech and Peterson 2003.

229. taking life unnecessarily: McIntyre 1995.

CHAPTER 8

236. toward wolf management: Musiani, Boitani, and Paquet 2009.

236. "symbolic wolf": Lawrence 1993.

237. these diverse groups: Hampton 1997; Ratti et al. 2004.

237. "hunting prowess, and power": Fivecrows 2007, x.

237. "connected with the wolf": Axtell quoted in Aurand 2010, 49.

237. "illness and possibly death": Fritts et al. 2003, 292.

237. base of totem poles: Joe Martin (Wickaninnish Island, British Columbia), interview, June 2011.

237. Olympic Peninsula, respectively: Ratti et al. 2004.

237. as the winter ceremonial: Ernst 1952.

237. some First Nations: Lewis George (Tofino, British Columbia), interview, June 2011.

237. "wolf is chosen": Ernst 1952, 48.

238. Changer named K'wati: Morganroth III 2002.

238. resurrected in 2007: Dickerson 2011.

238. 1800s and 1900s: Hampton 1997; Boitani 1995.

239. similarly positive light: Fritts et al. 2003.

239. in Europe and Asia: Linnell et al. 2002.

239. quantity of corpses: Hampton 1997.

239. "remnant populations": Hampton 1997.

239. to North America: Hampton 1997.

240. "early human ecology types": Boitani 1995.

240. still exists today: Boitani 2003.

240. ethnic group, Latinos: U.S. Census Bureau 2010.

241. "moral and naturalistic significance": Kellert 1996, 57, 110.

242. "federal tax dollars": Dan Warnock (Wallowa County, Oregon), interview, 8 August 2011.

242. from private sources: Suzanne Stone, telephone interview, 4 November 2011.

242. "is in a zoo?": McCormick quoted in Linton 1999, 1.

243. "things in perspective": Joel Reid (Mazama, Washington), email correspondence, 14 March 2011.

244. origins in western civilization: Nash 2001.

244. "who does not remain": Callicott 1994, 259.

245. with the natural world: Enns quoted in Lawson and Lawson 2006.

245. our own human nature: Pinkola Estes 2006.

245. hate or fear them: Lopez 1978.

246. dangerous to people: Treves and Karanth 2003.

246. Canada, Alaska: McNay 2002a.

246. and worldwide: Linnell et al. 2002.

246. than wild wolves do: Linnell et al. 2002.

246. "harmless wolf": Geist 2009, 28.

246. well-documented incident: McNay 2002a.

247. aggressive wolf encounters: Windle 2003, Theberge 2007.

247. "connection with nature": Wade 2005, 42.

251. "is very small": McNay 2002a, 837.

252. the incident on Vargas Island: McNay 2002b.

252. on this beach: Windle 2003.

252. food-conditioned wolves: Wang and Tedford 2008.

253. left the area: Sabina Mense (coordinator, Friends of Cortes Island citizen wolf monitoring program, Cortes Island, British Columbia), interviews, July 2010.

CHAPTER 9

259. side of the Olympic Mountains: WDFW 2011.

259. Cascades in 1946: ODFW 2010.

262. "ate the sheep": Richmond quoted in Dratch 1978.

262. "land had to provide": Anderson quoted in Dratch 1978.

262. "before crossing and going on": Murie notes from 1916–1917.

262. in the Elwah River valley: Scheffer 1995.

262. of the Quinault River: Dalquest 1948.

262. into the 1930s: Scheffer 1995.

262. as the early 1950s: Ratti et al. 2004.

262. "about their extinction": Dalquest 1948, 233.

263. disappearance of wolves from Oregon: Young and Goldman 1944.

263. disconnected populations: Krebs 2009.

263. Olympics' elk population: Murie 1935.

263. in the late 1990s: Ratti et al. 2004.

263. once and for all: McIntyre 1935.

263. "areas currently occupied": Wydeven et al. 2009, 91.

266. on the forest floor: Happe 1993.

267. seven hundred pounds: Schwartz and Mitchell 1945.

267. clearings as key features: Schwartz and Mitchell 1945; Jenkins and Starkey 1982.

270. subsequent rebounds: Happe 1993.

270. WDFW estimates: WDFW 2004.

270. three thousand more individuals: Jenkins and Manly 2008.

270. sites for homesteading: Patti Happe (Port Angeles, Washington), telephone interview, 13 August 2010.

270. west of the Cascades: Ratti et al. 2004.

270. which they hunted: Agee 1993.

270. forb understory: McShea 2005; Happe 1993; Schreiner et al. 1996; Woodward et al. 1994; Stockton et al. 2005.

271. habitats to the Olympics: Darimont and Paquet 2000; Ballard et al. 2001.

271. impact elk populations: Hebert et al. 1982; Quayle and Brunt 2003.

271. around the world: Ray et al. 2005; Terborgh and Estes 2010.
271. parts of the ecosystem: Jedrzejewska and Jedrzejewski 2005.
271. wolf numbers dropped: McLaren and Peterson 1994.
271. around the globe: McShea 2005.
271. deer and wolves: Stockton et al. 2005.
272. ungulate population increased: Beschta and Ripple 2008.
273. short-lived forest environment: Eisenberg 2010.
273. elk in some locations: Ratti et al. 2004.
273. change their behavior: Berger 2010.
273. threat of predation: Ripple and Beschta 2004.
273. predation by wolves there: Kauffman, Brodie, and Jules 2010.
274. research and debate: Kauffman, Brodie, and Jules 2010.
274. danger, or "vigilance": Liley and Creel 2008.
274. in Yellowstone elk: Wolff and Van Horn 2003.
276. diversity was decreased: Hebblewhite et al. 2005.
277. four and a half years of age: Suzanne Griffin, email correspondence, 11 November 2011.
277. a particular summer: Griffin et al. 2008.
277. own breeding territory: Griffin et al. 2009; Griffin 2007.
278. any other predator: Griffin et al. 2008.
278. carnivore scat analysis: Witczuk 2007.
278. for Olympic marmots: Griffin 2007.
278. limited and shrinking range: Griffin et al. 2008.
279. decades of the 1900s: Scheffer 1995; Dalquest 1948.
279. "lumbering and clearing proceed": Scheffer 1995, 74.
279. wolves became scarce: Dalquest 1948.
282. large food items: Bekoff and Wells 1986.

282. which they forage: Roemer, Gompper, and Van Valkenburgh 2009.
284. shared food sources: Buskirk 1999.
284. wolf-killed carcasses: Crabtree and Sheldon 1999.
284. often inversely related: Ballard, Carbyn, and Smith 2003.
284. ungulate carcasses: Wilmers, Crabtree, et al. 2003; Wilmers, Stahler, et al. 2003.
285. than coyotes do: Arjo, Pletscher, and Ream 2002.
285. around the globe: Ray, Redford, and Steneck 2005.
285. captive breeding program: Don Doyle (chair, Vancouver Island Marmot Recovery Team, Marmot Recovery Foundation), email correspondence, 1 November 2011.
285. another 21 percent: Bryant and Page 2005.
286. new, open habitat: Bryant and Page 2005.
286. these two large carnivores: Bryant and Page 2005.
286. in some locations: Don Doyle, email correspondence, 1 November 2011.
286. natural meadow habitat: Bryant and Janz 1996.
286. increasing deer populations: Hatter and Janz 1994.
286. distribution as well: Bryant and Janz 1996.
286. in ecological reserves: Don Doyle, email correspondence, 1 November 2011.
287. could be sustained there: Ratti 2004.
286. subsequently challenged: Mech 1996.
287. occupied as well: Wydeven et al. 2009.
287. the state into it: Wiles, Allen, and Hayes 2011.
287. in the state: Singleton, Gaines, and Lehmkuhl 2002.
293. retained fishing rights: Snoqualmie Indian Tribe 2011.

BIBLIOGRAPHY

Agee, J. K. 1993. *Fire Ecology of Pacific Northwest Forests.* Washington, D.C.: Island Press.

Almack, J. A., and S. H. Fitkin. 1998. Grizzly bear and gray wolf investigations in Washington State, 1994-1995. Final progress report. Washington Department of Fish and Wildlife, Olympia, Washington.

Apps, C. D., B. N. McLellan, T. A. Kinley, and J. P. Flaa. 2001. Scale-dependent habitat selection by mountain caribou, Columbia Mountains, British Columbia. *Journal of Wildlife Management* 65: 65-77.

Arjo, W. M., D. H. Pletscher, and R. R. Ream. 2002. Dietary overlap between wolves and coyotes in northwestern Montana. *Journal of Mammalogy* 83 (3): 754-766.

Armstrong, P. 2009. Conflict resolution and British Columbia's Great Bear Rainforest: lessons learned, 1995-2009. *Coast Forest Conservation Initiative.* http://www.coastforestconservationinitiative.com/pdf7/GBR_PDF.pdf.

Asa, Cheryl S., and L. David Mech. 1995. A review of the sensory organs in wolves and their importance to life history. In Carbyn, Fritts, and Seip 1995.

Aune, K. E., R. D. Mace, and D. W. Carney. 1994. The reproductive biology of female grizzly bears in the Northern Continental Divide ecosystem with supplemental data from the Yellowstone ecosystem. *International Conference on Bear Restoration and Management* 9 (1): 451-458.

Aurand, A. 2010. The wolves of the Wallowas. *1859: Oregon's Magazine* (Autumn): 44-51.

Ausband, D. E., M. S. Mitchell, K. Doherty, P. Zager, C. M. Mack, and J. Holyan. 2010. Surveying predicted rendezvous sites to monitor gray wolf populations. *Journal of Wildlife Management* 74: 1043-1049.

Ballard, W. B., L. N. Carbyn, and D. W. Smith. 2003. Wolf interactions with non-prey. In Mech and Boitani, eds. 2003.

Ballard, W. B., D. Lutz, T. W. Keegan, L. H. Carpenter, and J. C. deVos Jr. 2001. Deer-predator relationships: a review of recent North American studies with emphasis on mule and black-tailed deer. *Wildlife Society Bulletin* 29 (1): 99-115.

Bangs, E., M. Jimenez, C. Sime, S. Nadeau, and C. Mack. 2009. The art of wolf restoration in the northwestern United States. In Musiani, Boitani, and Paquet 2009.

Base, D. L., S. Zender, and D. Martorello. 2006. History, status, and hunter harvest of moose in Washington state. *Alces* 42: 111-114.

Bass, R. 2003. *The Ninemile Wolves.* New York: Mariner Books.

Bekoff, M., ed. 1978. *Coyotes: Biology, Behavior, and Management.* Caldwell, New Jersey: Blackburn Press.

Bekoff, M., and M. C. Wells. 1986. Social ecology and behavior of coyotes. *Advances in the Study of Behavior* 16: 251-338.

Berger, J. 2010. Fear-mediated food webs. In Terborgh and Estes 2010.

Berger, J., and D. W. Smith. 2005. Restoring functionality in Yellowstone with recovering carnivores: gains and uncertainties. In Ray, Redford, and Steneck 2005.

Berger, J., J. E. Swenson, and I. Persson. 2001. Recolonizing carnivores and naïve prey conservation lessons from Pleistocene extinctions. *Science* 291: 1036-1039.

Berger, J., P. B. Stacey, L. Bellis, and M. P. Johnson. 2001. A mammalian predator-prey imbalance: grizzly bear and wolf extinction affects avian neotropical migrants. *Ecological Applications* 11 (4): 947-960.

Berger, K. M. 2006. Carnivore-livestock conflicts; effects of subsidized predator control and economic correlates on the sheep industry. *Conservation Biology* 20 (3): 751-761.

Bergerud, A.T., and J. P. Elliot. 1986. Dynamics of caribou and wolves in northern British Columbia. *Canadian Journal of Zoology* 64: 1515-1529.

Bergerud, A. T., and R. E. Page. 1987. Displacement and dispersion of parturient caribou at calving as antipredator tactics. *Canadian Journal of Zoology* 67 (7): 1597-1606.

Beschta, R. L. 2003. Cottonwoods, elk, and wolves in the Lamar Valley of Yellowstone National Park. *Ecological Applications* 13 (5): 1295-1309.

Beschta, R. L., and W. J. Ripple. 2008. Wolves, trophic cascades, and rivers in the Olympic National Park, USA. *Ecohydrology* 1: 118-130.

Blackburn, T. M., P. Cassey, R. P. Duncan, K. L. Evans, and K. J. Gaston. 2004. Avian extinction and mammalian introductions on ocean islands. *Science* 305 (5692): 1955-1958.

Boitani, L. 1995. Ecological and cultural diversities in the evolution of wolf-human relationships. In Carbyn, Fritts, and Seip 1995.

———. 2003. Wolf conservation and recovery. In Mech and Boitani, eds. 2003.

Boitani, L., and P. Ciucci. 2009. Wolf management across Europe: species conservation without boundaries. In Musiani, Boitani, and Paquet 2009.

Bowman, J., J. C. Ray, A. J. Magoun, D. S. Johnson, and F. N. Dawson. 2010. Roads, logging, and the large-mammal community of an eastern Canadian boreal forest. *Canadian Journal of Zoology* 88: 454-467.

Boyd, D. K., R. R. Ream, D. H. Pletscher, and M. W. Fairchild. 1994. Prey taken by colonizing wolves and hunters in the Glacier National Park area. *Journal of Wildlife Management* 58 (2): 289-295.

Bradley, E. H., and D. H. Pletscher. 2005. Assessing factors related to wolf depredation of cattle in fenced pastures in Montana and Idaho. *Wildlife Society Bulletin* 33 (4): 1256-1265.

British Columbia Ministry of Agriculture and Lands. 2007. Unique collaboration to recover mountain caribou. News release, 2007AL0050-001308, 16 October. http://www.env.gov.bc.ca/wld/speciesconservation/mc/index.html.

British Columbia Ministry of Environment. Species at Risk Coordination. 2009. A Review of Management Actions to Recover Mountain Caribou in British Columbia. http://www.env.gov.bc.ca/wld/.../mc/files/Final_ST_Report_23Feb10.pdf.

Brewster, W. G., and S. H. Fritts. 1995. Taxonomy and genetics of the gray wolf in western North America: a review. In Carbyn, Fritts, and Seip 1995.

Bryant, A. A., and D. W. Janz. 1996. Distribution and abundance of Vancouver Island marmots (*Marmota vancouverensis*). *Canadian Journal of Zoology* 74: 667-677.

Bryant, A. A., and R. E. Page. 2005. Timing and causes of mortality in the endangered Vancouver Island marmot (*Marmota vancouverensis*). *Canadian Journal of Zoology* 83: 674-682

Bunnell, R. L., and A. C. Chan-McLeod. 1997. Terrestrial vertebrates. In *The Rain Forests of Home: Profile of a North American Bioregion*, edited by P. K. Schoonmaker, B. von Hagen, and E. C. Wolf. Washington, D.C.: Island Press.

Buskirk, S. W. 1999. Mesocarnivores of Yellowstone. In *Carnivores in Ecosystems: The Yellowstone Experience*, edited by T. W. Clark, A. P. Curlee, S. C. Minta, and P. M. Kareiva. New Haven: Yale University Press.

Callicott, J. B. 1994. The wilderness idea revisited: the sustainable development alternative. In *Reflecting on Nature*, edited by L. Gruen and D. Jamieson. New York: Oxford University Press.

Campbell, J. 2009. Report inconclusive; rancher said he saw wolves near dead cow. *Methow Valley News Online*, June 3. http://www.methowvalleynews.com/story.php?id=1487.

———. 2010. Wolf pelt investigation yields other wildlife charges for two Twisp men. *Methow Valley News Online*, April 7. http://methowvalleynews.com/story.php?id=3305.

Carbone, C., G. M. Mace, S. C. Roberts, and D. W. Macdonald. 1999. Energetic constraints on the diet of terrestrial carnivores. *Nature* 402: 286–288

Carbyn, L. N., S. H. Fritts, and D. R. Seip, eds. 1995. *Ecology and Conservation of Wolves in a Changing World.* Occasional Publication No. 35. Edmonton, Alberta: Canadian Circumpolar Institute, University of Alberta.

Carrera, R., W. Ballard, P. Gipson, B. T. Kelly, P. R. Krausman, M. C. Wallace, C. Villalobos, and D. B. Wester. 2008. Comparison of Mexican wolf and coyote diets in Arizona and New Mexico. *Journal of Wildlife Management* 72 (2): 376–381.

Carroll, C. 2007. Application of habitat models to wolf recovery planning in Washington. Unpublished report.

Carroll, C., M. K. Phillips, C. A. Lopez-Gonzalez, and N. H. Schumaker. 2006. Defining recovery goals and strategies for endangered species: the wolf as a case study. *BioScience* 56: 25–37.

Carroll, C., R. F. Noss, N. H. Schumaker, and P. C. Paquet. 2001. Is the return of the wolf, wolverine, and grizzly bear to Oregon and California biologically feasible? In *Large Mammal Restoration: Ecological and Sociological Challenges in the 21st Century*, edited by D. S. Maehr, R. F. Noss, and J. L. Larkin. Washington, D.C.: Island Press.

Centers for Disease Control and Prevention. 2011. Parasites—Echinococcosis. http://www.cdc.gov/parasites/echinococcosis.

Clavero, M., and E. Garcia-Berthou. 2005. Invasive species are a leading cause of animal extinctions. *TRENDS in Ecology and Evolution* 20 (3): 110.

Cockle, R. 2011. The debate over Oregon wolves spills into what to call them: gray wolves or Canadian gray wolves. *The Oregonian*, January 16. http://www.oregonlive.com/pacific-northwest-news/index.ssf/2011/01/the_debate_over_oregon_wolves_spills_into_what_to_call_them_gray_wolves_or_canadian_gray_wolves.html.

Colville National Forest. 2008. Colville Nation Forest Plan 1988: Amendments. http://www.fs.usda.gov/Internet/FSE_DOCUMENTS/fsbdev3_034850.doc.

Coppinger, R., L. Spector, and L. Miller. 2010. What, if anything, is a wolf? In Musiani et al. 2010.

Crabtree, R. L., and J. W. Sheldon. 1999. Coyotes and canid coexistence in Yellowstone. In *Carnivores in Ecosystems: The Yellowstone Experience*, edited by T. W. Clark, A. P. Curlee, S. C. Minta, and P. M. Kareiva. New Haven: Yale University Press.

Crandall, J. 2011. Letter: Wolves an economic drain on region. *Wallowa County Chieftain*, 28 April. http://www.wallowa.com/opinion/letters_to_editor/letter-wolves-an-economic-drain-on-region/article_a5cd105e-71c5-11e0-b886-001cc4c03286.html.

Creel, S., J. Winnie Jr., B. Maxwell, K. Hamlin, and M. Creel. 2005. Elk alter habitat selection as an antipredator response to wolves. *Ecology* 86: 3387–3397.

Csuti, B., T. O'Neil, M. Shaughnessy, E. Gaines, and J. Hak. 2001. *Atlas of Oregon Wildlife: Distribution, Habitat, and Natural History*. 2nd ed. Corvallis, Oregon: Oregon State University Press.

Dalquest, W. W. 1948. *Mammals of Washington*. Lawrence, Kansas: University of Kansas.

Darimont, C. T., and P. C. Paquet. 2000. *The Gray Wolves* (Canis lupus) *of British Columbia's Coastal Rainforests: Findings from Year 2000 Pilot Study and Conservation Assessment*. Prepared for the Raincoast Conservation Society, Victoria, BC.

Darimont, C. T., and P. C. Paquet. 2002. The gray wolves, *Canis lupus*, of British Columbia's central and north coast: distribution and conservation assessment. *Canadian Field Naturalist*. 116 (3): 416-422.

Darimont, C. T., P. C. Paquet, and T. E. Reimchen. 2008. Spawning salmon disrupt trophic coupling between wolves and ungulate prey in coastal British Columbia. *BMC Ecology* 8: 14.

———. 2009. Landscape heterogeneity and marine subsidy generate extensive intrapopulation niche diversity in a large terrestrial vertebrate. *Journal of Animal Ecology* 78 (1): 126-133.

Darimont, C. T., T. E. Reimchen, and P. C. Paquet. 2003. Foraging behaviour by gray wolves on salmon streams in coastal British Columbia. *Canadian Journal of Zoology* 81: 349-353.

Darimont, C. T., M. H. H. Price, N. N. Winchester, J. Gordon-Walker, and P. C. Paquet. 2004. Predators in natural fragments: foraging ecology of wolves in British Columbia's central and north coast archipelago. *Journal of Biogeography* 31: 1867-1877.

DellaSala, D., ed. 2010. *Temperate and Boreal Rainforests of the World: Ecology and Conservation*. Washington, D.C.: Island Press.

Deur, D., and N. J. Turner, eds. 2005. *Keeping It Living: Traditions of Plant Use and Cultivation on the Northwest Coast of North America*. Seattle: University of Washington Press.

Dickerson, P. 2011. Quileute welcome return of whales in ceremony. *Peninsula Daily News* (22 April). http://peninsuladailynews.com/article/20110422/NEWS/304229991.

Dittman, A. H., and T. P. Quinn. 1996. Homing in Pacific salmon: mechanisms and ecological basis. *Journal of Experimental Biology* 199: 83-91.

Donahue, D. L. 1999. *The Western Range Revisited: Removing Livestock from Public Lands to Conserve Native Biodiversity*. Norman, Oklahoma: University of Oklahoma Press.

Dratch, P. 1978. Interview transcriptions with Olympic Peninsula Old-Timers. Unpublished. Archive, Department of Cultural Resources, Olympic National Park, Port Angeles, Washington.

Dunlap, T. R. 1990. *Saving America's Wildlife: Ecology and the American Mind, 1850–1990.* Princeton, New Jersey: Princeton University Press.

Edwards, D. N. 2005. Carnivore-visitor use patterns within the Long Beach Unit of Pacific Rim National Park Reserve on the west coast of Vancouver Island. Parks Canada. http://www.clayoquotbiosphere.org/wildcoast/docs/4_Danielle_Edwards_Carnivore-Human_Interactions_in_Relation_to_Patterns_of_Human_Use.pdf.

Eisenberg, C. 2010. *The Wolf's Tooth: Keystone Predators, Trophic Cascades, and Biodiversity.* Washington, D.C.: Island Press.

Elbroch, M. 2006. *Animal Skulls: A Guide to North American Species.* Mechanicsburg, Pennsylvania: Stackpole Press.

Engelstoft, C. 2007. Black-tailed deer ecology in and around Pacific Rim National Park Reserve. Prepared for Pacific Rim National Park. Alula Biological Consulting. http://www.clayoquotbiosphere.org/wildcoast/docs/11_Christian_EngelStoft_Blacktail_Deer_Ecology_in_and_around_PRNPR_2007.pdf.

Ernst, A. 1952. *The Wolf Ritual of the Northwest Coast.* Eugene, Oregon: University of Oregon Press.

Fascione, N. 2006. *Places for Wolves: A Blueprint for Restoration and Recovery in the Lower 48 States.* Washington, D.C.: Defenders of Wildlife.

FiveCrows, J. 2007. Introduction: I am of this land. In *Nez Perce Country*, by Alvin M. Josephy Jr. Lincoln, Nebraska: University of Nebraska Press.

Foreyt, W. J., M. L. Drew, M. Atkinson, and D. McCauley. 2009. *Echinococcus granulosus* in gray wolves and ungulates in Idaho and Montana, USA. *Journal of Wildlife Diseases* 45: 1208–1212.

Forman, R. T. T., D. Sperling, J. A. Bissonette, A. P. Clevenger, C. D. Cutshall, V. H. Dale, L. Fahrig, R. France, C. R. Goldman, K. Heanue, J. A. Jones, F. J. Swanson, T. Turrentine, and T. C. Winter. 2003. *Road Ecology: Science and Solutions.* Washington, D.C.: Island Press.

Foreman, D. 2004. *Rewilding North America: A Vision for Conservation in the 21st Century.* Washington, D.C.: Island Press.

Frame, P. F., H. D. Cluff, and D. S. Hik. 2007. Response of wolves to experimental disturbance at homesites. *Journal of Wildlife Management* 71 (2): 316–320.

Friends of Clayoquot Sound. 2011a. Forest and logging fact sheet: Clayoquot Sound—May 2011. Friends of Clayoquot Sound. http://www.focs.ca/logging/factsheet.asp.

———. 2011b. History of Logging in Clayoquot Sound. http://www.focs.ca/logging/history.asp.

Fritts, S. H. 1992. Wolf recovery in the state of Washington. *International Wolf* (Winter): 19–20.

Fritts, S. H., E. E. Bangs, J. A. Fontaine, W. G. Brewster, and J. F. Gore. 1995. Restoring wolves to the northern Rocky Mountains of the United States. In Carbyn, Fritts, and Seip 1995.

Fritts, S. H., R. O. Stephenson, R. D. Hayes, and L. Boitani. 2003. Wolves and humans. In Mech and Boitani, eds. 2003.

Fuller, T. K. 1989. Population dynamics of wolves in north-central Minnesota. *Wildlife Monographs* No. 105. Bethesda, Maryland: The Wildlife Society.

Fuller, T. K., L. D. Mech, and J. F. Cochrane. 2003. Wolf population dynamics. In Mech and Boitani, eds. 2003.

Galle, A., M. Collinge, and R. Engeman. 2009. Trends in summer coyote and wolf predation on sheep in Idaho during a period of wolf recovery. In *Proceedings of the 13th Wildlife Damage Management Conference.* Saratoga Springs, New York.

Geist, V. 1981. Behavior: adaptive strategies in mule deer. In *Mule and Black-Tailed Deer of North America*, edited by O. C. Wallmo. Lincoln, Nebraska: University of Nebraska Press.

———. 1982. Adaptive behavioral strategies. In Thomas and Toweill 1982.

———. 2009. Let's get real: beyond wolf advocacy, toward realistic policies for carnivore conservation. *Fair Chase* (Summer): 26–33.

Gittleman, J. L. 1989. Carnivore group living: comparative trends. In Gittleman, ed. 1989.

Gittleman, J. L., ed. 1989. *Carnivore Behavior, Ecology, and Evolution.* Ithaca, New York: Cornell University Press.

Gonzalez-Hernandez, M. P., E. E. Starkey, and J. Karchesy. 2000. Seasonal variation in concentrations of fiber, crude protein, and phenolic compounds in leaves of red alder (*Alnus rubra*): nutritional implications for cervids. *Journal of Chemical Ecology* 26 (1): 293–301.

Griffen, K. A., M. Hebblewhite, H. S. Robinson, P. Zager, S. M. Barber-Meyer, D. Christianson, S. Creel, N. C. Harris, M. A. Hurley, D. H. Jackson, B. K. Johnson, W. L. Myers, J. D. Raithel, M. Schlegel, B. L. Smith, C. White, and P. J. White. 2011. Neonatal mortality of elk driven by climate, predator, phenology and predator community composition. *Journal of Animal Ecology* 80 (6): 1246–1257.

Griffin, S. C. 2007. Demography and ecology of a declining endemic: the Olympic marmot. Ph.D. thesis, University of Montana.

Griffin, S. C., P. C. Griffin, M. L. Taper, and L. S. Mills. 2009. Marmots on the move? Dispersal in a declining montane mammal. *Journal of Mammalogy* 90 (3): 686-695.

Griffin, S. C., M. L. Taper, R. Hoffman, and L. S. Mills. 2008. The case of the missing marmots: are metapopulation dynamics or range-wide declines responsible? *Biological Conservation* 141: 1293-1309.

Griffin, S. C., M. L. Taper, and L. S. Mills. 2007. Female Olympic marmots (*Marmota olympus*) reproduce in consecutive years. *American Midland Naturalist* 158: 221-225.

Groen, C. 2010. Op-ed: Lolo zone in perspective. Idaho Fish and Game Headquarters News Release, 8 March. Boise, Idaho. http://fishandgame.idaho.gov/public/media/viewNewsRelease.cfm?newsID=5339.

Halfpenny, J. C. 2003. *Yellowstone Wolves in the Wild*. Helena, Montana: Riverbend Publishing.

Hamashige, H. 2006. Dog virus may be killing Yellowstone wolves. *National Geographic News*, January 17. http://news.nationalgeographic.com/news/pf/22308220.html.

Hampton, B. 1997. *The Great American Wolf*. New York: Henry Holt and Company.

Hansen, H. J. 1986. *Wolves of Northern Idaho and Northeastern Washington*. Montana Cooperative Wildlife Research Unit, U.S. Fish and Wildlife Service.

Happe, P. 1993. Ecological relationships between cervid herbivory and understory vegetation in old-growth Sitka spruce-western hemlock forests in western Washington. Ph.D. thesis, Oregon State University.

Happe, P. J., K. J. Jenkins, E. E. Starkey, and S. H. Sharrow. 1990. Nutritional quality and tannin astringency of browse in clear-cuts and old-growth forests. *Journal of Wildfire Management* 54 (4): 557-566.

Hatler, D. F., D. W. Nagorsen, and A. M. Beal. 2008. *Carnivores of British Columbia*. Victoria, British Columbia: Royal British Columbia Museum.

Hatter, I. W., and D. W. Janz. 1994. Apparent demographic changes in black-tailed deer associated with wolf control on northern Vancouver Island. *Canadian Journal of Zoology* 72: 878-884.

Headley, S. A., D. G. Scorpio, O. Vidotto, and J. S. Dumier. 2011. *Neorickettsia helminthoeca* and salmon poisoning disease: a review. *The Veterinary Journal* 187 (2): 165-173.

Hebblewhite, M., D. H. Pletscher, and P. C. Paquet. 2002. Elk population dynamics in areas with and without predation by recolonizing wolves in Banff National Park, Alberta. *Canadian Journal of Zoology* 80: 789-799.

Hebblewhite, M., E. H. Merrill, and T. L. McDonald. 2005. Spatial decomposition of predation risk using resource selection functions: an example in a wolf-elk predator-prey system. *Oikos* 111: 101-111.

Hebblewhite, M., C. A. White, C. G. Nietvelt, J. A. McKenzie, T. E. Hurd, J. M. Fryxell, S. E. Bayley, and P. C. Paquet. 2005. Human activity mediates a trophic cascade caused by wolves. *Ecology* 86: 2135-2144.

Hebert, D. M., J. Youds, R. Davies, H. Langin, D. Janz, and G. W. Smith. 1982. Preliminary investigations of the Vancouver Island wolf (*Canis lupus crassodon*) prey relationships. In *Wolves of the World: Perspectives of Behavior, Ecology and Conservation*, edited by F. H. Harrington and P. C. Paquet. Park Ridge, New Jersey: Noyes Publications.

Hoekstra, J. M., T. M. Boucher, T. H Ricketts, and C. Roberts. 2005. Confronting a biome crisis: global disparities of habitat loss and protection. *Ecology Letters* 8: 23-29.

Holyan, J., K. Holder, J. Cronce, and C. Mack. 2011. *Wolf Conservation and Management in Idaho: Progress Report 2010*. Nez Perce Tribe Wolf Recovery Project, P.O. Box 365, Lapwai, Idaho; Idaho Department of Fish and Game, Boise, Idaho.

Howell, A. B. 1944. *Speed in Animals: Their Specialization for Running and Leaping*. Chicago: University of Chicago Press.

Hubbard, J. 2011. Letter: End of ranching. *Wallowa County Chieftain*, 28 July. http://wallowa.com/opinion/letters_to_editor/letter-end-of-ranching/article_5a5d8a62-b95b-11e0-a43d-001cc4c002e0.html.

Idaho Department of Fish and Game (IDFG). 2009. Wolves delisted: Idaho perspective. fishandgame.idaho.gov/public/docs/wolves/delistIdahoPersp.pdf.

———. 2010. Idaho Rule 10(j) Proposal, Lolo Zone. September 24. Boise, Idaho.

———. 2012. Wolf harvest 2011-2012. Updated May 18. http://fishandgame.idaho.gov/public/hunt/?getPage=121.

Idaho Department of Fish and Game and Nez Perce Tribe. 2012. *2011 Idaho Wolf Monitoring Progress Report*. Idaho Department of Fish and Game, Boise, Idaho; Nez Perce Tribe Wolf Recovery Project, P.O. Box 365, Lapwai, Idaho.

Idaho Legislature. 2011. Fiscal Year 2011, Legislative Budget Book: Department of Fish and Game. http://legislature.idaho.gov/budget/publications/PDFs/LBB/FY2011/NatRes/FishGameLBB.pdf.

Janis, C., J. Theodor, and B. Boisvert. 2002. Locomotor evolution in camels revisited: a quantitative analysis of pedal anatomy and the acquisition of the pacing gait. *Journal of Vertebrate Paleontology* 22 (1): 110-121.

Jedrzejewska, B., and W. Jedrzejewski. 2005. Large carnivores and ungulates in European temperate forest ecosystems: bottom-up and top-down control. In Ray et al. 2005.

Jeffers, R. 2002. *The Selected Poetry of Robinson Jeffers*. Edited by T. Hunt. Palo Alto, California: Stanford University Press.

Jenkins, K. J., and B. F. J. Manly. 2008. A double-observer method for reducing bias in faecal pellet surveys of forest ungulates. *Journal of Applied Ecology* 45: 1339-1348.

Jenkins, K. J., and E. E. Starkey. 1982. Social organization of Roosevelt elk in an old-growth forest. *Journal of Mammalogy* 63 (2): 331-334.

Johnson, R. E., and K. M. Cassidy. 1997. *Terrestrial Mammals of Washington State: Location Data and Predicted Distributions*. Volume 3 in *Washington State Gap Analysis Project Final Report*, edited by K. M. Cassidy, C. E. Grue, M. R. Smith, and K. M. Dvornich. Seattle: Washington Cooperative Fish and Wildlife Research Unit, University of Washington.

Josephy Jr., A. M. 2007. *Nez Perce Country*. Lincoln, Nebraska: University of Nebraska Press.

Kauffman, J. B., M. Mahrt, L. A. Mahrt, and W. D. Edge. 2001. Wildlife of riparian habitats. In *Wildlife-Habitat Relationships in Oregon and Washington*, edited by D. H. Johnson and T. A. O'Neil. Corvallis, Oregon: Oregon State University Press.

Kauffman, M. J., J. F. Brodie, and E. S. Jules. 2010. Are wolves saving Yellowstone's aspen? A landscape-level test of a behaviorally mediated trophic cascade. *Ecology* 91: 2742-2755.

Kauffman, M. J., N. Varley, D. W. Smith, D. R. Stahler, D. R. MacNulty, and M. S. Boyce. 2007. Landscape heterogeneity shapes predation in a newly restored predator-prey system. *Ecological Letters* 10 (8): 690-700.

Kay, C. E. 1994. Aboriginal overkill—the role of Native-Americans in structuring western ecosystems. *Human Nature* 5 (4): 359-398.

Kellert, S R. 1996. *The Value of Life: Biological Diversity and Human Society*. Washington, D.C.: Island Press.

Kertson B. N. 2010. Cougar ecology, behavior, and interactions with people in a wildland-urban environment in western Washington. Ph.D. thesis, University of Washington.

Kinley, T. A., and C. D. Apps. 2001. Mortality patterns in a subpopulation of endangered mountain caribou. *Wildlife Society Bulletin* 29 (1): 158-164.

Kohira, M., and E. A. Rexstad. 1997. Diets of wolves, *Canis lupus*, in logged and unlogged forests of southeastern Alaska. *Canadian Field-Naturalist* 111 (3): 429-435.

Krebs, C. J. 2009. *Ecology: The Experimental Analysis of Distribution and Abundance*. 6th ed. San Francisco: Pearson Education, Inc.

Kreeger, T. J. 2003. The internal wolf: physiology, pathology, and pharmacology. In Mech and Boitani, eds. 2003.

Krkosek, M., J. S. Ford, A. Morton, S. Lele, R. A. Myers, and M. A. Lewis. 2007. Declining wild salmon populations in relation to parasites from farm salmon. *Science* 318: 1772-1775.

Kunkel, K., and D. H. Pletscher. 2001. Winter hunting patterns of wolves in and near Glacier National Park, Montana. *Journal of Wildlife Management* 65 (3): 520-530.

Kunkel, K. E., D. H. Pletscher, D. K. Boyd, R. R. Ream, and M. W. Fairchild. 2004. Factors correlated with foraging behavior of wolves in and near Glacier National Park, Montana. *Journal of Wildlife Management* 68 (1): 167-178.

Kunkel, K. E., T. K. Ruth, D. H. Pletscher, and M. G. Hornocker. 1999. Winter prey selection by wolves and cougars in and near Glacier National Park, Montana. *Journal of Wildlife Management* 63 (3): 901-910.

Lange, I. M. 2002. *Ice Age Mammals of North America: A Guide to the Big, the Hairy, and the Bizarre.* Missoula, Montana: Mountain Press.

Larsen, T., and W. J. Ripple. 2006. Modeling gray wolf (*Canis lupus*) habitat in the Pacific Northwest, U.S.A. *Journal of Conservation Planning* 2: 17-33.

Laufer, J. R., and P. T. Jenkins. 1989. *A Preliminary Study of the Grey Wolf History and Status in the Region of the Cascade Mountains of Washington State.* Tenino, Washington: Wolf Haven America.

Lawrence, E. A. 1993. The sacred bee, the filthy pig, and the bat out of hell: animal symbolism as cognitive biophilia. In *The Biophilia Hypothesis*, edited by S. R. Kellert and E. O. Wilson, 301-344. Washington, D.C.: Island Press.

Lawson, S., and S. Lawson. 2006. Wolf Project Based on Traditional Knowledge. Tofino, British Columbia. Unpublished proposal.

Leakey, R., and R. Lewin. 1995. *The Sixth Extinction: Patterns of Life and the Future of Humankind.* New York: Doubleday.

Liebenberg, L. 1990. *The Art of Tracking: The Origin of Science.* Claremount, South Africa: David Philip Publishers Ltd.

Liley, S., and S. Creel. 2008. What best explains vigilance in elk: characteristics of prey, predators, or the environment. *Behavioral Ecology* 19 (2): 245-254.

Linnell, J. D. C., R. Andersen, L. Balciauskas, J. C. Blanco, L. Boitani, S. Brainerd, U. Beitenmoser, I. Kojola, O. Liberg. J. Loe, H. Okarma, H. C. Pedersen, C. Promberger, H. Sand, E. J. Solberg, H. Valdmann, and P. Wabakken. 2002. The fear of wolves: a review of wolfs attacks on humans. *NINA Norsk institutt for naturforskning Oppdragsmelding* 731.

Linton, Darrell. 1999. *Crying Wolf: The Misuse of Public Opinion to Prevent Endangered Species Recovery*. Port Angeles, Washington: Peninsula Environmental Center.

Littell, J. S., M. McGuire Elsner, L. C. Whitely Binder, and A. K. Snover, eds. 2009. Executive summary for *The Washington Climate Change Impacts Assessment: Evaluating Washington's Future in a Changing Climate*. Seattle, Washington: Climate Impacts Group, University of Washington. www.cses.washington.edu/db/pdf/wacciaexecsummary638.pdf.

Lopez, B. 1978. *Of Wolves and Men*. New York, New York: Charles Scribner's Sons.

Mack, C., J. Rachael, J. Holyan, J. Husseman, M. Lucid, and B. Thomas. 2010. *Wolf Conservation and Management in Idaho: Progress Report 2009*. Nez Perce Tribe Wolf Recovery Project, P.O. Box 365, Lapwai, Idaho; Idaho Department of Fish and Game, Boise, Idaho.

Mahoney, S. 2004. The North American wildlife conservation model: triumph for man and nature. *Bugle* 21 (3). http://www.rmef.org/NewsandMedia/PubsTV/Bugle/2004/MayJune/Features/NAModel.html.

Marquis, R. J. 2010. The role of herbivores in terrestrial trophic cascades. In Terborgh and Estes 2010.

McAllister, I. 2007. *Last Wild Wolves: Ghosts of the Great Bear Rainforest*. With contributions from C. T. Darimont. Vancouver, British Columbia: Greystone Books.

McAllister, I., and N. Read. 2010. *The Sea Wolves: Living Wild in the Great Bear Rainforest*. Victoria, British Columbia: Orca Book Publishers.

McArthur, C., C. T. Robbins, A. E. Hagerman, and T. A. Hanley. 1993. Diet selection by a ruminant generalist browser in relation to plant chemistry. *Canadian Journal of Zoology* 71: 2236–2243.

McIntyre, R., ed. 1995. *War Against the Wolf: America's Campaign to Exterminate the Wolf*. Stillwater, Minnesota: Voyageur Press.

———. 1996. *A Society of Wolves: National Parks and the Battle over the Wolf*. Revised edition. Stillwater, Minnesota: Voyageur Press, Inc.

McLaren, B. E., and R. O. Peterson. 1994. Wolves, moose, and tree rings on Isle Royale (Isle Royale National Park, Michigan). *Science* 266 (5190): 1555–1558.

McNay, M. E. 2002a. Wolf-human interactions in Alaska and Canada: a review of the case history. *Wildlife Society Bulletin* 30 (3): 831–843.

———. 2002b. *A Case History of Wolf-Human Encounters in Alaska and Canada*. Wildlife Technical Bulletin 13. Alaska Department of Fish and Game, Division of Wildlife Conservation.

McNay, R. S., and J. M. Voller. 1995. Mortality causes and survival estimates for adult female Columbian black-tailed deer. *Journal of Wildlife Management* 59 (1): 138-146.

McShea, W. J. 2005. Forest ecosystems without carnivores: when ungulates rule the world. In Ray et al. 2005.

Mech, L. D. 1970. *The Wolf: The Ecology and Behavior of an Endangered Species.* Minneapolis: University of Minnesota Press.

———. 1974. *Canis lupus. Mammalian Species Account No. 37.* The American Society of Mammalogists.

———. 1996. Prediction failure of a wolf landscape model. *Wildlife Society Bulletin* 34 (3): 874-877.

———. 2007. Femur-marrow fat of white-tailed deer fawns killed by wolves. *Journal of Wildlife Management* 71 (3): 920-923.

Mech, L. D., and L. Boitani. 2003. Wolf social ecology. In Mech and Boitani, eds. 2003.

Mech, L. D., and L. Boitani, eds. 2003. *Wolves: Behavior, Ecology and Conservation.* Chicago: University of Chicago Press.

Mech, L. D., and R. O. Peterson. 2003. Wolf-prey relations. In Mech and Boitani, eds. 2003.

Meehl, G. A., T. F. Stocker, W. D. Collins, P. Friedlingstein, A. T. Gaye, J.M. Gregory, A. Kitoh, R. Knutti, J. M. Murphy, A. Noda, S. C. B. Raper, I. G. Watterson, A. J. Weaver, and Z.-C. Zhao. 2007. Global climate projections. In *Climate Change 2007: The Physical Science Basis. Contribution of Working Group I to the Fourth Assessment Report of the Intergovernmental Panel on Climate Change*, edited by S. Solomon, D. Qin, M. Manning, Z. Chen, M. Marquis, K. B. Averyt, M. Tignor, and H. L. Miller. Cambridge, UK, and New York: Cambridge University Press.

Meiri, S., and T. Dayan. 2003. On the validity of Bergmann's rule. *Journal of Biogeography* 30: 331-351.

Milne, D. G., A. S. Harestad, and K. Atkinson. 1989. Diets of wolves on northern Vancouver Island. *Northwest Science* 63 (3): 83-86.

Miquelle, D. G., P. A. Stephens, E. N. Smirnov, J. M. Goodrich, O. J. Zaumyslova, and A. E. Myslenkov. 2005. Tigers and wolves in the Russian Far East: competitive exclusion, functional redundancy, and conservation implications. In Ray et al. 2005.

Morgan, R. 2012. *Oregon Wolf Conservation and Management Plan: 2011 Annual Report.* Oregon Department of Fish and Wildlife, La Grande, Oregon.

Morganroth III, C. 2002. Quileute. In *Native Peoples of the Olympic Peninsula: Who We Are*, edited by J. Wray. Norman, Oklahoma: University of Oklahoma Press.

Muhly, T. B., M. Alexander, M. S. Boyce, R. Creasey, M. Hebblewhite, D. Paton, J. A. Pitt, and M. Musiani. 2010. Differential risk effects of wolves on wild versus domestic prey have consequences for conservation. *Oikos* 119 (8): 1243-1254.

Muhly, T. B., and M. Musiani. 2009. Livestock depredation by wolves and the ranching economy in the Northwestern United States. *Ecological Economics* 68: 2439-2450.

Muñoz-Fuentes, V., C. T. Darimont, R. K. Wayne, P. C. Paquet, and J. A. Leonard. 2009a. Ecological factors drive differentiation in wolves from British Columbia. *Journal of Biogeography* 36 (8): 1516-1531.

Muñoz-Fuentes, V., C. T. Darimont, P. C. Paquet, and J. A. Leonard. 2009b. The genetic legacy of extirpation and re-colonization in Vancouver Island wolves. *Conservation Genetics* 11 (2): 547-556.

Murie, A. 1935. Special report of senior naturalist technician Adolf Murie on wildlife of the Olympics. Department of the Interior, National Park Service, Wildlife Division, Washington, D.C., cited in Ratti et al. 2004.

Murie, O. Notes from 1916-1917 field work in the Elwah River Valley. Unpublished. Olympic National Park Archives.

Musiani, M., L. Boitani, and P. C. Paquet, eds. 2009. *A New Era for Wolves and People: Wolf Recovery, Human Attitudes, and Policy.* Calgary: University of Calgary Press.

Musiani, M., L. Boitani, and P. C. Paquet, eds. 2010. *The World of Wolves: New Perspectives on Ecology, Behavior and Management.* Calgary, Alberta: University of Calgary Press.

Nash, R. 2001. *Wilderness and the American Mind.* 4th edition. New Haven: Yale University Press.

National Research Council, Committee on Protection and Management of Pacific Northwest Anadromous Salmonids. 1996. *Upstream: Salmon and Society in the Pacific Northwest.* Washington, D.C.: National Academy Press.

Nelson, M. P., J. A. Vucetich, P. C. Paquet, and J. K. Bump. 2011. An inadequate construct? North American model: what's flawed, what's missing, what's needed. *The Wildlife Professional* (Summer): 58-60.

Nietvelt, C. G. 2001. Herbivory interactions between beaver (*Castor canadensis*) and elk (*Cervus elaphus*) on willow (*Salix* spp.) in Banff National Park, Alberta. Master's thesis, University of Alberta.

Nowak, R. M. 2003. Wolf evolution and taxonomy. In Mech and Boitani, eds. 2003.

Oakleaf, J. K., C. Mack, and D. L. Murray. 2003. Effects of wolves on livestock calf survival and movements in central Idaho. *Journal of Wildlife Management* 67 (2): 299-306.

Oakleaf, J. K., D. L. Murray, J. R. Oakleaf, E. E. Bangs, C. M. Mack, D. W. Smith, J. A. Fontaine, M. D. Jimenez, T. J. Meier, and C. C. Niemeyer. 2006. Habitat selection by recolonizing wolves in the Northern Rocky Mountains of the United States. *Journal of Wildlife Management* 70 (2): 554-563.

Oftedal, O. T., and J. L. Gittleman. 1989. Patterns of energy output during reproduction in carnivores. In Gittleman, ed. 1989.

Okanogan County District Court. 2009. Affidavit for Search Warrant. Case No. 09-0150. Okanogan, Washington.

Olson, R., and W. Hubert. 1994. *Beaver: Water Resources and Riparian Habitat Manager.* Laramie, Wyoming: University of Wyoming.

O'Neil, T. A., K. A. Bettinger, M. Vander Heyden, B. G. Marcot, C. Barret, T. K. Mellen, W. M. Vanderhaegen, D. H. Johnson, P. J. Doran, L. Wunder, and K. M. Boula. 2001. Structural conditions and habitat elements of Oregon and Washington. *In Wildlife-Habitat Relationships in Oregon and Washington*, edited by D. H. Johnson and T. A. O'Neil. Corvallis, Oregon: Oregon State University Press.

Oregon Department of Fish and Wildlife (ODFW). 2003. Oregon's elk management plan, February 2003. Oregon Department of Fish and Wildlife. Portland, Oregon.

———. 2009a. Oregon wolf program April–May 2009 update. Oregon Department of Fish and Wildlife, Salem, Oregon. http://www.dfw.state.or.us/wolves/docs/oregon_wolf_program/2009_april_wolf_report.pdf.

———. 2009b. Oregon wolf program August–September 2009 update. Oregon Department of Fish and Wildlife, Salem, Oregon. http://www.dfw.state.or.us/wolves/docs/oregon_wolf_program/2009_august-september_wolf_update.pdf.

———. 2010. Oregon wolf conservation and management plan. Oregon Department of Fish and Wildlife, Salem, Oregon.

———. 2011. Imnaha pack timeline of events. http://www.dfw.state.or.us/Wolves/imnaha_wolf_pack.asp.

Orlando, A. 2009. Mountain caribou recovery plan implementation falls short. *Revelstoke Times Review*, 23 February. http://mountaincaribou.ca/news/100.

Packard, J. M. 2003. Wolf behavior: reproductive, social, and intelligent. In Mech and Boitani, eds. 2003.

Palomares, F., and T. M. Caro. 1999. Interspecific killing among mammalian carnivores. *The American Naturalist* 153 (5): 492-508

Paquet, P. C., S. Alexander, S. Doneln, and C. Callaghan. 2010. Influence of anthropogenically modified snow conditions on wolf predatory behaviour. In Musiani, Boitani, and Paquet 2010.

Person, D. K., M. K. Kirchhoff, V. Van Ballenberghe, G. C. Iverson, and E. Grossman. 1996. The Alexander Archipelago wolf: a conservation assessment. General technical report. PNW-GTR-384. Department of Agriculture, USDA Forest Service, Pacific Northwest Research Station, Portland, Oregon.

Peterson, R. O., and P. Ciucci. 2003. The wolf as carnivore. In Mech and Boitani, eds. 2003.

Pimentel, D., L. Westra, and R. Noss. 2000. *Ecological Integrity: Integrating Environment, Conservation, and Health.* Washington, D.C.: Island Press.

Pinkola Estes, C. 2006. *Women Who Run With the Wolves: Myths and Stories of the Wild Woman Archetype.* New York: Ballantine Books.

Prater, Y. 1983. *Snoqualmie Pass: From Indian Trail to Interstate.* Seattle, Washington: Mountaineers Books.

Price, M. H. H., C. T. Darimont, N. N. Winchester, and P. C. Paquet. 2005. Facts from faeces: prey remains in wolf, *Canis lupus,* faeces revise occurrence records for mammals of British Columbia's coastal archipelago. *Canadian Field Naturalist* 119 (2): 192–196

Quayle, J. F., and K. R. Brunt 2003. *Status of Roosevelt elk* (Cervus elaphus roosevelti) *in British Columbia.* Wildlife Bulletin No. B-106. B.C. Ministry of Water, Land and Air Protection, Biodiversity Branch, Victoria, British Columbia.

Rachael, J., ed. 2010. *Elk.* Project W-170-R-34. Progress Report. July 1, 2009–June 30, 2010. Idaho Department of Fish and Game, Boise, Idaho.

Raincoast Conservation Foundation. 2010. *What's at Stake? The Cost of Oil on British Columbia's Priceless Coast.* Sidney, British Columbia: Raincoast Conservation Foundation.

Ratti, J. T., M. Weinstein, J. M. Scott, P. A. Wiseman, A. Gillesberg, C. A. Miller, M. M. Szepanski, and L. K. Svancara. 2004. Feasibility of wolf reintroductions to Olympic Peninsula, Washington. *Northwest Science* 78 (Special Issue).

Ray, J. C., K. H. Redford, R. S. Steneck, and J. Berger, eds. 2005. *Large Carnivores and the Conservation of Biodiversity.* Washington, D.C.: Island Press.

Reimchen, T. E. 2000. Some ecological and evolutionary aspects of bear-salmon interactions in coastal British Columbia. *Canadian Journal of Zoology* 78: 448–457.

Remington, T. 2009. Editor's note. Two-thirds of Idaho wolf carcasses examined have thousands of hydatid disease tapeworms, by George Dovel. *Black Bear Blog.* http://www.skinnymoose.com/bbb/2010/01/06/two-thirds-of-idaho-wolf-carcasses-examined-have-thousands-of-hydatid-disease-tapeworms/.

Riggs, R. A., A. R. Tiedemann, J. G. Cook, T. M. Ballard, P. J. Edgerton, M. Vavra, W. C. Krueger, F. C. Hall, L. D. Bryant, L. L. Irwin, and T. Delcurto. 2000. Modification of mixed-conifer forests by ruminant herbivores in the Blue Mountains Ecological Province. USDA Forest Service, Pacific Northwest Research Station. PNW-RP-527.

Ripple, W. J., and R. L. Beschta. 2003. Wolf reintroductions, predation risk, and cottonwood recovery in Yellowstone National Park. *Forest Ecology and Management* 184: 299-313.

———. 2004. Wolves and the ecology of fear: can predation risk structure ecosystems? *Bioscience* 54 (8): 755-766.

Ripple, W. J., T. P. Rooney, and R. L. Beschta. 2010. Large predators, deer, and trophic cascades in boreal and temperate ecosystems. In Terborgh and Estes 2010.

Roemer, G. W., M. E. Gompper, and B. Van Valkenburgh. 2009. The ecological role of the mammalian mesocarnivore. *BioScience* 59 (2): 165-173.

Routledge, R. 2001. Mixed-stock vs. terminal fisheries: a bioeconomic model. *Natural Resource Modeling* 14: 523-539.

Scott, J. W., and R. L. De Lorme. 1988. *Historical Atlas of Washington*. Norman, Oklahoma: University of Oklahoma Press.

Scheffer, V. B. 1995. *Mammals of the Olympic National Park and Vicinity*. Northwest Fauna Occasional Monographs on Vertebrate Natural History Number 2. Olympia, Washington: Society for Northwestern Vertebrate Biology.

Schoen, J. W. 1972. Mammals of the San Juan Archipelago: distribution and colonization of native land mammals and insularity in three populations of *Peromyscus maniculatus*. Master's thesis, University of Puget Sound.

Schreiner, E. G., K. A. Krueger, P. J. Happe, and D. B. Houston. 1996. Understory patch dynamics and ungulate herbivory in old-growth forests of Olympic National Park, Washington. *Canadian Journal of Forestry Resources* 26: 255-265.

Schwartz, J. E., and G. E. Mitchell. 1945. The Roosevelt elk on the Olympic Peninsula, Washington. *Journal of Wildlife Management* 9 (4): 295-319.

Scott, B. M.V., and D. M. Shackleton. 1982. A preliminary study of the social organization of the Vancouver Island wolf (*Canis lupus crassodon*; Hall, 1932). In *Wolves of the World: Perspectives of Behavior, Ecology and Conservation*, edited by F. H. Harrington and P. C. Paquet. Park Ridge, New Jersey: Noyes Publications. 12-25.

Seip, D. R. 1992. Factors limiting woodland caribou populations and their interrelationships with wolves and moose in southeastern British Columbia. *Canadian Journal of Zoology* 70: 1494-1503.

Singleton, P. H., W. L. Gaines, and J. F. Lehmkuhl. 2002. Landscape permeability for large carnivores in Washington: a geographic information system weighted-distance and least-cost corridor assessment. Research Paper PNW-RP-549, Pacific Northwest Research Station, USDA Forest Service, Portland, Oregon.

Smith, D. W., R. O. Peterson, and D. B. Houston. 2003. Yellowstone after wolves. *BioScience* 53 (4): 330-340.

Smith, D. W., T. D. Drummer, K. M. Murphy, D. S. Guernsey, and S. B. Evans. 2004. Winter prey selection and estimation of wolf kill rates in Yellowstone National Park, 1995-2000. *Journal of Wildlife Management* 68: 153-166.

Smith, K. G., E. J. Ficht, D. Hobson, T. C. Sorensen, and D. Hervieux. 2000. Winter distribution of woodland caribou in relation to clear-cut logging in west-central Alberta. *Canadian Journal of Zoology* 78 (8): 1433-1440.

Snoqualmie Indian Tribe. 2011. Signatories to the 1855 Treaty of Port Elliot. http://www.snoqualmienation.com/about/about.htm.

Stahler, D. R. 2000. Interspecific interactions between the common raven (*Corvus corax*) and the gray wolf (*Canis lupus*) in Yellowstone National Park, Wyoming: investigations of a predator and scavenger relationship. Master's thesis, University of Vermont. Cited in Peterson and Ciucci 2003.

Stahler, D. R., D. W. Smith, and D. S. Guernsey. 2006. Foraging and feeding ecology of the gray wolf (*Canis lupus*): lessons from Yellowstone National Park, Wyoming, USA. *The Journal of Nutrition* 136 (7 Suppl): 1923S-1926S.

Stockton, S. A., S. Allombert, A. J. Gaston, and J. L. Martin. 2005. A natural experiment on the effects of high deer densities on the native flora of coastal temperate rainforests. *Biological Conservation* 126: 118-128

Stone, S., N. Fascione, C. Miller, J. Pissot, G. Schrader, and J. Timberlake. 2008. *Livestock and Wolves: A Guide to Nonlethal Tools and Methods to Reduce Conflicts.* Washington, D.C.: Defenders of Wildlife.

Stotyn, S. A., B. N. McLellan, and R. Serrouya. 2007. *Mortality Sources and Spatial Partitioning among Mountain Caribou, Moose, and Wolves in the North Columbia Mountains, British Columbia.* Nelson, British Columbia: Columbia Basin Fish and Wildlife Compensation Program.

Suttles, W., and K. Ames. 1997. Pre-European history. In *The Rainforests of Home: Profile of a North American Bioregion*, edited by P. K. Schoonmaker, B. von Hagen, and E. C. Wolf. Washington, D.C.: Island Press.

Szepanski, M., M. Ben-David, and V. Van Ballenberghe. 1999. Assessment of anadromous salmon resources in the diet of the Alexander Archipelago wolf using stable isotope analysis. *Oecologia* 120: 327-335.

Terborgh, J., and J. A. Estes. 2010. *Trophic Cascades: Predators, Prey and the Changing Dynamics of Nature*. Washington, D.C.: Island Press.

Theberge, M. 2007. Human encounters with wolves and cougars in the Pacific Rim National Park Reserve area: summary and analysis of behaviour. Wild Coat Project, Pacific Rim National Park. http://www.clayoquotbiosphere.org/wildcoast/docs/5a_Michelle_Theberge_Wolf_Cougar%20Encounters_%20Mar29_07.pdf.

Thomas, J. A., M. G. Telfer, D. B. Roy, C. D. Preston, J. J. D. Greenwood, J. Asher, R. Fox, R. T. Clarke, and J. H. Lawton. 2004. Comparative losses of British butterflies, birds, and plants and the global extinction crisis. *Science* 303 (5665): 1879-1881.

Thomas, J. W., and D. E. Toweill. 1982. *Elk of North America: Ecology and Management*. Harrisburg, Pennsylvania: Stackpole Books.

Toweill, D. E., and G. Vecellio. 2004. Shiras moose in Idaho: status and management. *Alces* 40: 33-43.

Trapp, J. 2004. Wolf den site selection and characteristics in the northern Rocky Mountains: a multi-scale analysis. Master's thesis, Prescott College.

Treves, A., and K. U. Karanth. 2003. Human-carnivore conflict and perspectives on carnivore management worldwide. *Conservation Biology* 17 (6): 1491-1499.

Two gray wolf packs discovered in northern Washington. 1990. *Endangered Species Technical Bulletin* 15 (6): 6.

Ulanowicz, Robert E. 2000. Toward the measurement of ecological integrity. In Pimentel et al. 2000.

U.S. Census Bureau. 2010. State and County Quick Facts. http://quickfacts.census.gov/qfd/states/53000.html.

U.S. District Court 2011. Case 3:11-cv-00023-JO. District of Oregon, Portland Division. Filed January 6, 2011.

U.S. Fish and Wildlife Service, Nez Perce Tribe, National Park Service, Montana Fish, Wildlife and Parks, Blackfeet Nation, Confederated Salish and Kootenai Tribes, Idaho Fish and Game, and USDA Wildlife Services. 2008. Rocky Mountain wolf recovery 2007 interagency annual report. Edited by C. A. Sime and E. E. Bangs. U.S. Fish and Wildlife Service, Helena, Montana.

Verts, B. J., and L. N. Carraway. 1998. *Land Mammals of Oregon*. Berkeley, California: University of California Press.

Vucetich, J. A., R. O. Peterson, and T. A. Waite. 2004. Raven scavenging favours group foraging in wolves. *Animal Behavior* 67: 1117-1126.

Wade, S. 2005. Visitor behaviour and perception of bears, wolves and cougars at Pacific Rim National Park Reserve. B.Sc. honors project. University of Victoria. http://www.clayoquotbiosphere.org/wildcoast/docs/7_Sasha_Wade_Visitor_Perception_of_Bears_Wolves_and_Cougars_at_PRNPR_2005.pdf.

Wang, X., and R. H. Tedford. 2008. *Dogs: Their Fossil Relatives and Evolutionary History*. New York: Columbia University Press.

Washington Department of Fish and Wildlife (WDFW). 2004. Olympic elk herd plan. Wildlife Program, Washington Department of Fish and Wildlife, Olympia, Washington.

———. 2011. Final environmental impact statement for the wolf conservation and management plan for Washington. Washington Department of Fish and Wildlife, Olympia, Washington.

———. 2012a. Grey wolf conservation and management: frequently asked questions. http://wdfw.wa.gov/conservation/gray_wolf/faq.html#8.

———. 2012b. Wildlife managers treat dead calf as 'probable' case of wolf predation. News release, 23 May. http://wdfw.wa.gov/news/release-print/may2312b/.

Wayne, R. K. 2010. Recent advances in the population genetics of wolf-like canids. In Musiani, Boitani, and Paquet 2010.

Westra, L., P. Miller, J. R. Karr, W. E. Rees, and R. E. Ulanowicz. 2000. Ecological integrity and the aims of the Global Integrity Project. In Pimentel et al. 2000.

Wildcoast 2010. Learning to live with large carnivores: Wildcoast Project Primer and Guideline. EKOS Communications Inc. and Parks Canada Agency. http://ekoscommunications.com/files/wildcoast/Final_WildCoast_Primer_V3.pdf.

Wildlife Services 2011a. PDR Table A. Wildlife services, fiscal year 2010, federal and cooperative funding by resource. U.S. Department of Agriculture, Washington, D.C. http://www.aphis.usda.gov/wildlife_damage/prog_data/2010_prog_data/index.shtml.

Wildlife Services 2011b. Table G Short Report. Animals taken by Wildlife Services-FY2010. U.S. Department of Agriculture, Washington, D.C. http://www.aphis.usda.gov/wildlife_damage/prog_data/2010_prog_data/index.shtml.

Wiles, G. J., H. L. Allen, and G. E. Hayes. 2011. Wolf conservation and management plan for Washington. Washington Department of Fish and Wildlife, Olympia, Washington.

Williams, E. S., and I. K. Barker, eds. 2001. *Infectious Diseases of Wild Mammals*. 3rd ed. Ames, Iowa: Iowa State University Press.

Wilmers, C. C., R. L. Crabtree, D. W. Smith, K. M. Murphy, and W. M. Getz. 2003. Trophic facilitation by introduced top predators: grey wolf subsidies to scavengers in Yellowstone National Park. *Journal of Animal Ecology* 72: 909-916.

Wilmers, C. C., D. R. Stahler, R. L. Crabtree, D. W. Smith, and W. M. Getz. 2003. Resource dispersion and consumer dominance: scavenging at wolf and hunter-killed carcasses in Greater Yellowstone, USA. *Ecological Letters* 6: 996-1003.

Windle, T. 2003. A case history of wolf-human encounters in and around Pacific Rim National Park Reserve 1983-2003. University of Northern British Columbia. http://www.clayoquotbiosphere.org/wildcoast/docs/2_Todd_Windle_Wolf-Human_Pet_Interaction_1983_2003.pdf.

Wipfli, M. S., J. P. Hudson, and J. P. Caouette. 2003. Marine subsidies in freshwater ecosystems: salmon carcasses increase the growth rate of stream-resident salmonids. *Transactions of the American Fisheries Society* 132: 371-381.

Witczuk, J. J. 2007. Monitoring program and assessment of coyote predation for Olympic marmots. Master's thesis, University of Montana.

Wittmer, H. U., B. N. McLellan, D. R. Seip, J. A. Young, T. A. Kinley, G. S. Watts, and D. Hamilton. Population dynamics of the endangered mountain ecotype of woodland caribou (*Rangifer tarandus caribou*) in British Columbia, Canada. *Canadian Journal of Zoology* 83: 407-418.

Wolf, E. C., A. P. Mitchell, and P. K. Schoonmaker. 1995. *The Rainforests of Home: An Atlas of People and Place.* Part 1: Natural Forests and Native Languages of the Coastal Temperate Rain Forest. Edited by E. L. Kellogg. Portland, Oregon: Ecotrust, Pacific GIS, and Conservation International.

Wolff, J. O., and T. Van Horn. 2003. Vigilance and foraging patterns of American elk during the rut in habitats with and without predators. *Canadian Journal of Zoology* 81 (2): 266-271.

Woodroffe, R., and J. R. Ginsberg. 2005. King of the beasts? Evidence for guild redundancy among large mammalian carnivores. In Ray, Redford, and Steneck 2005.

Woodward, A., E. G. Schreiner, D. B. Houston, and B. B. Moorhead. 1994. Ungulate-forest relationships on the Olympic Peninsula: retrospective exclosure studies. *Northwest Science* 68: 97-110.

Wydeven, A. P., R. L. Jurewicz, T. R. Van Deelen, J. Erb, J. H. Hammill, D. E. Beyer Jr., B. Roell, J. E. Wiedenhoeft, and D. A. Weitz. 2009. Gray wolf conservation in the Great Lakes region of the United States. In Musiani, Boitani, and Paquet 2009.

Yochim, M. J. 2001. Aboriginal overkill overstated errors in Charles Kay's hypothesis. *Human Nature* 12 (2): 141-167.

Young, S. P., and E. A. Goldman. 1944. *Wolves of North America.* New York: Dover Publications.

ACKNOWLEDGMENTS

This book is a product of the deep love, respect, and devotion that defines my marriage to Darcy Ottey. Without her unwavering support, this endeavor would not have progressed from its origins as a sketchy idea discussed while eating bacon and drinking red wine huddled by a woodstove in a cabin in northwestern Montana on a snowy winter day. The tenets of our marriage—to support each other in following our passions and bringing beauty into this world through them—were tested on many levels over the span of this book's creation but her support for this vision did not falter. That she also agreed to be the lead author on the chapter in her field of specialty didn't hurt either. For all of it I am grateful.

I am grateful to my parents, Johanna Goldfarb and Ralph Moskowitz, for all the many opportunities they have given me in this life in general and the support and encouragement that each of them gave me in numerous large and small ways for this project in particular.

Several individuals provided invaluable help with my fieldwork, research, and writing. Foremost, Emily Gibson joined me on numerous field trips across the region, assisted me with literature research, reviewed the entire manuscript (most of it more than once), and helped me clarify how to present key parts of the story. Her combination of passion for the subject matter along with her understanding of wildlife biology concepts, her skill as wildlife tracker and naturalist, and her keen feedback were essential to this project. Similarly, Brian McConnell's support included joining me for fieldwork and lending his tremendous expertise with wildlife to the project. His perspectives as a conservationist and lifelong hunter were greatly valued. Steve and Susanne Lawson were amazingly generous and helpful with my time in Clayoquot Sound. I gained a great deal from the time I spent in the field with them and am grateful for their perspectives and decades of experience with wolves and conservation. Cristina Eisenberg provided numerous research opportunities for me as well as ideas and inspiration for the subject matter in the book. Thomas Murphy was tremendously helpful in designing the approach Darcy and I took to writing the chapter on wolves and human culture and providing resources along the way. Thanks to Mark Darrach for being my guide, interpreter, and mentor of all things botanical both in the field and out. I am grateful to Roger Bean for his assistance in a variety of ways, all relating to Canada. Thanks also to Leah Gerrard, Drew Middlebrooks, Brandon Sheely, Gabe Spence, Rob Nagel, and Marcus Reynerson.

A number of professionals, organizations, and agencies graciously allowed me to join them for field trips and created time for interviews, and assisted in various ways with the research of this book. Ray Robertson, a man who wears many hats related to wolves in Washington, is foremost in my mind in this regard. From the Washington Department of Fish and Wildlife I am grateful for the support of Jay Shepherd, Paul Frame, Scott Fitkin, Gary Wiles, William Moore, and Harriet Allen. From the Oregon Department of Fish and Wildlife I would like to thank Russ Morgan. Thanks much to John Rohrer, Michael Borysewicz, and Terry Reynolds of the U.S. Forest Service. Conservation Northwest provided support and inspiration for the project. In particular, I am grateful to Jasmine Minbashian, George Wooten, Jen Watkins, Paul Bannick, and Erin Moore. Lisa Lauf Rooper, Kenyon Fields, and other staff at Wildlands Network provided support in fundraising efforts and feedback in designing the content of the book. I am grateful to the hospitality and field assistance provided by Raincoast Conservation Foundation and their field station manager, Douglas Brown. Chris Darimont, their science director, shared his time generously for an interview and ongoing correspondence through the project. The numerous articles and published reports by him and others at Raincoast were similarly an amazing asset. Thanks to Greg Dyson and Jennifer Schwartz at Hells Canyon Preservation Council, Bob Hanson of Pacific Rim National Park, and Sabina Mense from Friends of Cortes Island. Thanks to the Heiltsuk Nation for providing me with a research permit on short notice and for allowing me to collect material for the book in their traditional territory. Analisa Fenix of Ecotrust was exceptional in her efforts to produce accurate and beautiful maps for the book. Thanks to Jenn Wolfe for producing wolf and coyote silhouettes.

More than seventy individuals provided financial backing for this project. Without their support, it would never have gotten off the ground. I would specifically like to thank Ralph Moskowitz, Mallory Clarke and the Cascade Tracking Team, Johanna Goldfarb and Kerry Levin, Kate Thayer, Mark and Heather Timken, Alisa Malloch and her hiking friends, Kim and Chris Chisholm of Wolf Camp, Ken Hackett, Bob and Janet Conklin, Joan Golston, Brian McConnell, and Rob Speiden of Natural Awareness Tracking School. I am very grateful for in-kind donations and support from several organizations. Raincoast Conservation Foundation donated the use of field resources. My supervisors, colleagues, and students at Wilderness Awareness School were patient and supportive with my unusual and challenging schedule and at times divided attention, as well as providing research opportunities associated with my teaching schedule. Lindsey Swope donated the use of the lodge at Skalitude Retreat Center in the Methow Valley and the Seattle Patagonia store donated clothing and other equipment for my fieldwork.

I drew a great deal of inspiration from the following writers, photographers, and conservationists: Karston Heuer, Amy Gulick, Paul Bannick, Cristina Eisenberg, Doug Chadwick, and, in particular, Pacific Wild founder Ian McAllister, whose books were an inspiration for me in starting this project. Ian's use of stories and images to connect people with wildlands and conservation have deeply informed my own efforts here. On a more practical level, Toby Malloy and Kristi Dranginis provided technical support for my photography.

I am exceptionally grateful for the patience, coaching, and mentorship that Eve Goodman, managing editorial director at Timber Press, has provided throughout this project, from helping to define the scope of the book and reviewing early drafts to helping me find my voice as a narrative writer. I have learned a tremendous amount about writing from the time and energy she and Timber Press have invested in me and this book. Laken Wright brought her creative vision to the beautiful design and layout.

I am grateful to the following individuals for reviewing one or more chapters and offering critical feedback on them: Ray Robertson, Chris Darimont, Jay Shepherd, Greg Dyson, Patti Happe, Tom Murphy, Mark Elbroch, Michael Borysewicz, Paul Frame, Cristina Eisenberg, Sue Griffin, Jennifer Schwartz, Don Doyle, Brandon Sheely, Mark Darrach, Steve Nadeau, Ross Wilson, Gabe Spence, Steve Smith, and Bob Hansen. Thanks also to Tamara Walker, Emily Gibson, and Brian McConnell for reviewing the entire manuscript.

Of course in any endeavor of this size, completion would be impossible without the help of many. My research for this book included reading thousands of pages of scientific articles and popular literature, interviewing numerous experts in related fields, listening to the stories of people involved in topics related to those of this book, and hundreds of days in the field attempting to track down and observe wolves across the region with the help and guidance of many people. Even with all the support of so many talented people, I am sure mistakes will be found, and for these I am solely responsible.

DAVID MOSKOWITZ

Carnation, Washington

INDEX

Ahousaht First Nation, 235
Alaska, 16, 55, 85, 153
alder, 85, 219, 266, 272-73, 292
Almack, John, 44
American marten (*Martes americana*), 282
Anderson, Charlie, 262
Arctic fox (*Vulpes lagopus*), 251
aspen, 46, 96, 102, 221-23, 275, 276
Axtell, Horace, 237

badger, 97
balsam fir (*Abies balsamea*), 271
balsamroot, 46, 221
Banff National Park, 128, 178, 275-76
barren ground caribou (*Rangifer tarandus groenlandicus*), 93
bears, 48, 63, 72, 85, 97, 98, 139, 205. See also black bear; grizzly bear
    caching food, 115
    co-opting carcasses, 96-98, 117, 120
    elk and, 212-13
    human contact, 120, 252
    salmon and, 141, 148, 153
    wolves and, 160-62
Bear Valley pack, 208-15
beaver (*Castor canadensis*), 35, 113, 114, 128-32, 154, 177, 219-20, 276
    riparian forest and, 129, 215, 220, 275
Bella Bella, British Columbia, 133, 144, 145, 146
Bermann, Christian, 55
bighorn sheep (*Ovis canadensis*), 160, 175
bison (*Bison bison*), 175
bitterbrush, 46, 221
black bear (*Ursus americanus*), 40, 46, 55, 69, 101, 103, 139, 157, 161, 224, 226

attack by, 246
feeding behavior and diet, 40-41, 46, 78-79, 97, 139, 205, 266
wolves and, 98, 40-41, 101, 160-62
black-tailed deer (*Odocoileus hemionus columbianus*), 40, 125, 154, 157, 159, 271
black tree lichens (*Byoria* species), 93, 94
Blue Mountains, 15, 196-229
    browsing pressure in, 205-8, 219, 266
    Sinks, 206
    timber industry in, 202-3, 206
    wildlife in, 202, 205
    wolf packs in, 201-4, 206
    wolves and elk in, 205-16
    wolves and livestock in, 202-3, 216-29
bobcats, 160, 224
Boitani, Luigi, 240
boreal chickadees, 86
Borysewicz, Michael, 92-93
British Columbia, 16, 35, 138, 141, 144
    caribou decline and, 95
    Central Coast, 134-56
    coyotes and, 279
    moose in, 92
    oil pipeline and, 163
    sea wolves, 144
    timber industry and, 143
    wolf diet in, 37, 144, 148-53
    wolf diet studies, 130, 132
    wolf DNA, 34, 178, 182
    wolf extirpation, 159
    wolf habitat studies, 184
    wolf reestablishment, 178
    wolves in, 34, 52-53, 56, 58, 73, 75-76, 77, 92, 127, 136-37, 140, 142, 144, 173, 185, 271
Brown, Douglas, 145-46, 150, 156, 166

California, 16, 35, 259, 267
    wolf habitat in, 188
    wolf reestablishment in, 178
Canada, 44, 55, 87, 92. See also British Columbia
Canada lynx, 86
Canis *lepophagus*, 54, 173
caribou (*Rangifer tarandus*) or mountain caribou, 85, 86, 93, 94-95, 102, 184
    conservation, 95-96, 104-5
    habitat destruction by, 94-96, 125
    wolves and, 94-95, 286
carnivores, 54, 55, 62, 63, 65, 114. See also social carnivores, wolves as
    aggression by, 246
    apex, 22, 54, 173, 195, 271, 285
    competition for food and, 162, 252, 284
    density of prey and, 286
    ecological role, 215, 271, 285
    in Pacific Northwest, 55, 97, 176
    predator-prey interaction, 160-62, 226, 273
    prehistoric, 54, 172-73
    reproduction rates, 72
    resurgence, globally, 124
    social, 54, 116
    solitary, caching of food, 115
    wolf interaction with, 96-99
Cascade Crest, 30, 41
Cascade Range, 16, 19, 35, 221, 259
Cascades Citizen Wildlife Monitoring Project, 30-31, 33
ceanothus, 207
Chilcotin, 237
Clayoquot Sound, 64, 138, 157, 235
    wolf attack, 246-47
    wolves of, 138, 160-61, 164-65, 232-35

Clockum Ridge, 221–22
Coast Range, 35, 259
coloration, 57–59, 106, 140, 234
Columbia Highlands, 84, 86, 87, 104–5
    wolf return to, 97, 180
Columbia River, 16, 39, 153, 176, 201, 221, 259
Colville National Forest, 92–93, 102
coneflowers, 206
Confederated Tribes of the Umatilla, 206
conservation biology, 39
Conservation Northwest, 33
Cortes Island, 123, 132, 138, 253
cottonwood, 20, 206, 219, 272, 292
cougars, 48, 73, 123, 205, 261, 267, 286
coyote (Canis latrans), 54
    characteristics, 56–59, 62, 66
    diet, 282, 290–93
    distribution, 46, 260, 266, 283
    evolutionary history, 54
    expansion of, 279, 285
    extirpation of, 285
    food sources, 97, 132, 284
    as mesocarnivore, 278–85
    Olympic marmot and, 260, 276–78
    predator control and, 224
    as scavengers, 266, 280, 284
    wolves and, 56–57, 279, 280, 282, 284–85
crows, 139, 140, 141, 292

Dalquest, Walter, 262
Darimont, Chris, 73
Darrach, Mark, 205–6
deer, 94, 154, 266. See also black-tailed deer; mule deer
    carnivores and behavior, 36, 49
    deforestation and, 154
    hunting of, 177
    as prey, 39, 40, 47, 63, 92, 93, 103, 123, 124, 126, 127, 128, 132, 154, 209
    seasonal migration, 39, 44, 78, 91
    wolf pack size and, 75

Defenders of Wildlife, 204, 215, 241, 242
dens/den sites, 45, 46–47, 64, 65, 70, 77, 158, 159, 160, 186–87
Diamond pack, 84, 87–92, 106, 112, 113, 127
    home range of, 77, 87–90, 102, 182–83, 185
diet, 18, 128. See also prey
    adaptability, 18, 132, 282
    beaver in, 114, 128–32, 154
    black bears in, 40–41, 161
    bones and marrow, 63, 113
    caching food, 64, 70, 115
    carcass consumption, 40–41, 116–17, 118–19
    as carnivorous, 63, 78, 282
    on Clayoquot Sound, 160–61
    of coastal wolves, 127, 144
    deforestation and, 154
    digestive system and, 70
    feeding behavior, 41
    garbage, 114, 133
    grizzly bear in, 98, 132
    large hoofed animals as primary, 54, 113, 120, 128, 154, 183
    livestock in, 227–29
    marine food sources, 114, 117, 127, 132, 144, 153
    marmot in, 285–86
    omnivorous aspects, 54, 63, 114, 132
    opportunistic, 114, 128–33
    raccoon in, 130, 139
    regurgitation and, 70, 115
    in Rockies, 128
    salmon in, 36, 37, 49, 114, 117, 127, 132, 145, 148–54, 163, 166–67, 175, 183, 292
    salmon poisoning and, 152–53, 262–63
    scavenged carcasses, 114, 129, 133, 284
    skull and teeth and, 60–63
    small mammals/rodents, 114
    on Vancouver Island, 128
dire wolves (Canus dirus), 54–55, 173
dog (Canis lupus familiaris), 57, 62

Douglas fir, 46, 86, 103, 206, 221
Ducks Unlimited, 212

Eagle Cap Wilderness, 201, 202, 227
eagles, 224, 286, 292
ecosystems
    adjacent landscapes and connectivity, protecting, 39
    beavers and, 129, 215, 220, 275
    browsing pressure and overgrazing, 205–8, 265–73
    climate shifts and, 174
    dead trees in, 144
    European-American colonization and, 177
    in flux, 174
    humans in, 19, 37, 163
    livestock and, 216–29
    lower elevations as unprotected, 38–39
    mining and, 95, 174, 219
    North Cascades, 48–49
    predators' role, 48, 97, 99, 113, 117, 215
    rainforests, 16, 41, 141, 144, 265–73
    riparian, 20, 215, 219–21, 265–76
    road ecology, 194
    scavengers in, 117
    of Selkirks, 85–87
    songbirds and, 215, 275, 276
    timber industry and, 94–95, 96, 102, 104–5, 126, 154, 163, 182, 193, 286
    ungulate density and low biodiversity, 271, 275–76
    wolves and, 18, 19, 22, 37, 96, 97, 99, 113–14, 117, 124, 128, 144, 215–16, 271, 273, 275–76, 285
Eisenberg, Cristina, 96, 124, 127, 278–79
elk (Cervus elaphus), 19–20, 68, 222, 267. See also Roosevelt elk
    bachelor herds, 120
    bears as predators of, 212–13
    decline, causes, 205, 212–13

defensive strategies, 121, 123
diet, 19-20, 272
habitats, 39, 126-27, 214
history, 175
hunting of, 177, 202, 203, 205, 212-15, 267
landscape impacted by, 205-8, 222, 265-73, 274
Lolo herd, 213-14
Olympic Peninsula, 263-76
predator interaction, 274-76
as prey, 63, 92, 93, 96, 110-11, 118-19, 120, 124, 126-28, 132, 154, 201, 216-17, 267
riparian forest and, 20, 270-73
in Salmon River Mountains, 208-15
seasonal migration, 78, 91
subspecies, 267
wolf pack size and, 75
wolves and, 204, 205-16
Elwah River valley, 262
Engelmann spruce, 86
Enns, Eli, 244-45
Enterprise, Oregon, 201
Ernst, Alice, 237-38
Europe and British Isles, 194, 238-40, 263, 271
extinction event, 21
extinctions
    of *Canis lepophagus*, 55
    of dire wolf, 55
    elk, regional, 293
    human-caused, 124-25
    marmots as endangered, 286
    novel predators and invasive species, 124
    of Pacific Northwest wolf, 178
    Pleistocene epoch, 173
    wolf-caused unlikely, 124

First Nations people, 176, 177, 178
    elk hunting and, 270
    land and wildlife management, 240-41
    totem animals, 237
    Treaty of Port Elliot, 293
    Wolf Ritual, 237-38
    wolves as religious, cultural icons, 235-38, 260

Fitkin, Scott, 44
FiveCrows, Jeremy, 237
forest fires, 86, 102, 104-5, 207, 212, 214
Forks, Washington, 242
Frame, Paul, 41
Frank Church-River of No Return Wilderness, 208
Frasier River, 16, 34

Geist, Valerius, 246
Gibson, Emily, 102, 120, 216
Glacier National Park, 178, 180
Glacier Peak Wilderness, 44
Goldman, Edward, 262-63
Great Basin Desert, 16
Great Bear Rainforest, 52-53, 138, 141, 145-56, 158
    salmon and terminal fishery, 163, 166-67
    swimming wolves in, 162, 164-65
    wolf rendezvous site, 186-87
Great Lakes wolves, 263-64, 271, 286-87
Griffin, Sue, 276, 277, 278
grizzly bear (*Ursus arctos*), 44, 55, 85, 96-98, 100-101, 127, 141
    aggression by, 246
    extirpation of, 215, 224
    feeding behavior and diet, 78-79, 96-98, 100, 141
    recovery zone, Selkirks, 102, 104-5
    reproduction rates, 72
    salmon and, 161-62
    skull and teeth of, 62
    wolves and, 73, 96-98, 132, 160-62
Groen, Cal, 213

habitat. *See also* specific packs
    connectivity of landscapes and, 184, 287
    forested, 263-64
    future in Pacific Northwest, 182-85, 188-89
    genetics and, 182
    historical, 176
    home range sizes and, 35, 77
    human challenges and, 185, 191
    land developers and, 226-27

landscape driving wolf travel choices, 190-95
modern preferences, 176-77, 183, 188-89
Olympic Peninsula as, 263
prey and, 184
rainforest, 134-67, 170-71
roads and, 193, 194-95
southern portions of Northwest, 188
Hampton, Bruce, 239
Hansen, Bob, 161
Happe, Patti, 266-67
hawks, 224
hearing, sense of, 62
Heiltsuk First Nation, 146, 151, 166-67
    Fisheries Program, 166
Hells Canyon, 201
    Preservation Council, 220
Himalayan blackberry, 280-81
hoary marmot (*Marmota caligata*), 35
Hoh River, 265-66, 269, 270-75
howling, 30, 44, 106, 112, 153, 208, 209
Hozomeen Mountain, 40, 43
Hozomeen wolves, 40-45
huckleberries, 85, 132, 268-69
Hudson's Bay Co., 87, 177
Huettman, Lindsay, 274-75
hunting and hunters, 229, 272
    caribou decline and, 94-95
    of elk, 117, 202, 203, 205, 212-15, 267, 293
    of grizzlies, 293
    "lethal take" permit, 242
    Vancouver Island, 286
    of wolves, 44, 73, 87, 95, 145, 173, 185, 193, 194, 195, 264-65, 293
hunting behavior of wolves, 40, 77, 78-79, 114-17
    aggression vs. flight in prey and, 121
    areas preferred for kills, 40
    charging, 250-51
    chases, 99, 113, 114-15, 127
    classic hunt, 121
    feet and claws and, 64-65
    fishing, 150, 151-52
    habitat, terrain, and, 126-27, 184
hunting group size, 114

injuries and, 63, 92, 73, 152
  legs, locomotion, and speed,
    64-65
  location of prey, 114
  selection of prey, 120-25
  size of prey and, 117
  strategies for big game, 115,
    121, 123
  surplus killing, 228-29
  training of young in, 116
Idaho, 153, 201. *See also* Selkirk
    Mountains
  elk and wolves, 212-15
  wolf dispersal from, 201
  wolf extirpation in, 56
  wolf hunting, 15, 185
  wolf kills in, 120, 126
  wolf packs in, 208-15
  wolf reestablishment in, 56,
    87, 180, 208, 218
  wolf-tracking in, 126-27
  wolves in, 15, 56, 77, 153, 178,
    180, 185
Idaho Department of Fish and
    Wildlife (IDFW), 212-15
Imnaha pack, 201-4, 227-28
  home range of, 202
  livestock and, 216, 229
  wolf, B300, 201-2
Imnaha River, 198-99, 201
Isle Royal National Park, 271

Jasper National Park, 178
Jeffers, Robinson, 124
Joseph, Oregon, 201-4

Keating wolves, 225
Klamath-Siskiyou Mountains,
    259, 270

Lake Chelan, 39
  -Sawtooth Wilderness, 39,
    190
larch, 221
Last Wild Wolves, The (McAllis-
    ter), 161
Lawrence, Elizabeth, 236
Lawson, Steve and Susanne,
    157, 162
Liebenberg, Louis, 18
livestock, 39, 216-29
  anti-wolf sentiment and, 38,
    239, 240-42
  behavior and wolves, 227
  in Blue Mountains, 202, 203

depredation of, 48, 123, 185,
    225, 227-29, 239, 241-42,
    261-62, 293
  ecology and, 176, 219-20
  fear of wolves and, 204, 224
  Imnaha pack and, 202
  Keating wolves and, 225
  leasing of public lands and
    grazing permits, 202, 215,
    218-19, 220, 221, 225, 226
  Lookout pack and, 39, 48
  number killed by wolves vs.
    dogs, 228
  overgrazing by, 206-7,
    219-22
  predator control and, 222,
    224-27, 229
  Taylor Grazing Act and, 218
  in Washington State, 218
  wolf habitat and, 183
  wolves ignoring of, 123
lodgepole pines, 86
Lookout pack, 30-37, 38, 39,
    45-48, 74
  breeding female disappears,
    45-47
  den site, 45, 46-47
  DNA of, 34, 37, 182
  food sources, 35, 36, 39, 116,
    118
  home range, 32-33, 35, 38,
    39, 77, 87, 89, 185, 188
  livestock and, 37
  monitoring of, 34, 35, 36, 46,
    47
  number of wolves in, 35
  poaching of, 35, 37, 38,
    48-49, 75-76
  pups, 34

Maquinna (Lewis George),
    235-36, 237, 238
marbled murrelets, 141
marmot. *See* Olympic marmot;
    Vancouver Island marmot
Martin, Joe, 237
McAllister, Ian, 144, 161
McCadie, Malcolm, 276
McCarty, Jesse, 46, 48
McConnell, Brian, 128
McCormick, George, 242
McNay, Mark, 247
Mense, Sabina, 253
mesocarnivores, 278-85
Methow River, 31, 39

Methow Valley, 30, 39, 128
Mexican wolves, 55
mining industry, 95, 174, 219
mink, 153, 160
Minnesota, 263, 264
Montana, 96, 99
  hunting of wolves in, 185
  wolf diet in, 128
  wolf home range sizes, 77
  wolf hunting behavior
    study, 127
  wolves in, 58, 59, 71, 87, 118-
    19, 123, 185, 193
moose (*Alces alces*), 35, 85, 92
  behavior and wolves, 92, 124
  in Blue Mountains, 202
  hunting of, 177
  population increase, 182
  as prey, 77, 93, 103, 113, 114,
    123, 124, 185, 271
  riparian forest and, 215, 271
  size of, 93, 113
  timber industry and, 92, 94,
    185, 194
  tracks, 91-92
  wolf injuries and, 63, 123
  wolf pack size and, 75
Morse, Susan, 97
mountain ash, 85, 207
mountain goats (*Oreamnos
    americanus*), 175, 184
mountain lion (*Puma con-
    color*), 55, 78-79, 94, 98-99,
    160, 176, 205
  aggression by, 246
  caching food, 115
  physical characteristics, 62,
    63, 65, 66,
  predator control and, 224
  prey for, 93, 129, 132, 213
  wolves and, 97, 98-99
Mount Baker, 41, 44
Mount Lassen, 16
Mount Redoubt, 26-27
Mount Saint Helens, 188-89
Mount Shasta, 16
mule deer (*Odocoileus heminus*),
    35, 39, 46, 66, 99, 118, 201
Murie, Adolf, 263
Murie, Olaus, 262, 263, 267
muskrats, 160
mustelids, 63, 65
Nagel, Rob, 102
Neorickettsia helminthoeca, 152-
    53, 262

Nez Perce, 178, 218, 237
North Cascades, 24–49, 190
 anti-wolf sentiment and
  poachers, 35, 37, 38, 47–48
 black bear in, 40
 characteristics of, 34, 38, 48
 connectivity with other
  wildlands, 34–35, 39
 ecosystem, 48–49
 federally protected areas,
  29, 38
 grizzly recovery zone in,
  102
 map, 28
 Ross Lake in, 40–45
 salmon runs, 34
 springtime in, 32–33
 timber and cattle grazing,
  29
 wolf future in, 48–49
 wolf packs in, 30–37, 38, 39,
  45–48, 49
 wolf tracking in, 29–30
 wolves, return of, 30, 37
 wolves in, 29–49, 56, 182,
  243–44, 245
North Cascades National Park,
 48–49
Northeast Washington For-
 estry Coalition, 102
northern bog lemmings, 86
northern hawk owls, 86
North Fork Flathead River
 wolves, 84, 96, 99, 178
North Fork of the Burnt River,
 219–21, 226
Nuu-cha-nulth, 235, 237

O'Higgins, Mark Kang, 274–75
Okanogan Highlands, 35
Okanogan-Wenatchee
 National Forest, 36
Olympic marmot (Marmota
 olympus), 276–78, 285
Olympic National Park, 191,
 242, 256–57, 258, 260
 elk without wolves, 265–76
 return of wolves (proposed),
  263, 286–87
 Roosevelt elk and, 262, 264
Olympic Peninsula, 189, 191,
 259–78
 coyotes and, 276–79
 ecology and elk, 265–76
 elk of, 267–68, 270–73

First Nations of, 237, 238
 marmot endangered,
  276–78
 timber industry and, 260
 wolf disappears, 260–65
Oregon, 35, 188, 258, 259
 beavers in, 129
 fossilized wolf remains, 177
 livestock and riparian
  destruction, 219–21
 livestock and wolves,
  224–25
 OR-7, travels of, 189
 timber industry, 259
 wolf delisting as endan-
  gered, 15
 wolf extirpation, 201, 258
 wolf habitat, 182, 188, 189
 wolf home range sizes, 77
 wolf management plans,
  195
 wolf packs in, 201, 206,
  227–28
 wolf population, 15, 178
 wolves return to, 57, 178, 180,
  201
Oregon Department of Fish
 and Wildlife (ODFW),
 202, 203, 206, 241
Ottey, Darcy, 230, 234–35
Outward Bound, 190, 243

Pacific fisher (Martes pennanti), 282
Pacific Northwest, 16, 17, 222
 carnivores in, 55
 coyotes in, 276–85
 European arrival in, 238
 extirpation of wolves and
  grizzlies, 97, 178, 282, 293
 Ice Age in, 175
 land developers in, 226–27
 mountains of, 34–35
 prehistoric, 172
 public opinion about wolf
  protection, 37–38, 124
 rainforests, 16, 34, 41, 141,
  144
 reestablishment of wolves,
  grizzlies in, 16, 97, 124, 173
 restoration of historic forest in, 102
 wildlife of, 93, 113, 153, 175
 wolf as only large social carnivore in, 54, 55

wolf diet in, 114, 120
wolf distribution in, 175
wolf habitat, future, 182–85,
 188–89
wolf home range sizes, 77
wolf population in, 15
wolf reproduction in, 71, 72
wolf routes through, 190–95
Pacific silver fir, 41
Pacific yew, 206, 207
parvovirus, 73
Person, David, 73
poaching, 38, 47–49
 of Lookout pack, 35, 37, 38,
  48–49, 75–76, 188
poisoning of wolves, 95, 251,
 261–62, 282, 293
ponderosa pines, 41, 46, 86,
 221
population of wolves
 dispersal, 184–85, 189–93
 in Europe, 194
 expansion of, 178, 180
 future predictions, 183–85
 genetics and, 180, 182
 habitat and, 183
 historical, 177
 humans as defining factor,
  185, 195
 in Idaho, 15, 178
 lowest levels, 178
 map, 179
 in Oregon, 15, 178
 prey density and, 183
 reproductive rates and, 70,
  185
 sources of and dispersal,
  current, 178–80
 sources and sinks, 184–85
 translocations, 178, 180,
  287
 variables in, 183
 in Washington State, 15, 178
prey, 39. See also specific types
 accessibility of, 184
 coevolution with predators,
  124
 for Diamond pack, 92
 effect of carnivores on
  behavior, 35–36, 99, 114–
  15, 124, 273–76
 human-built environment
  and, 125, 127
 for Lookout pack, 35, 36, 39,
  116, 118

prey [*continued*]
  in low-elevation land-
    scapes, 39
  predator-prey dynamics,
    160–62, 226, 273, 274–76
  response to attack, aggres-
    sion vs. flight, 121
  for Rockies wolves, 35
  for Salmo pack, 103
  seasonal migration, 77, 78
  selection of, 120–25
  size of, 75, 78, 114
  vulnerability factors, 121,
    123
  wolf density and, 183, 263,
    286
  wolf teeth, feet, and, 63, 64
pronghorn sheep (*Antilocarpa
  americana*), 175
Puget Sound, 40, 122–23, 141,
  218, 253, 259, 260
  coyotes in, 278–81, 292, 293
  wolves in, 177, 183, 191, 237

Quileute Nation, 238
Quinault people, 237
Quinault River, 262

rabbits, 160, 286
rabies, 239, 251, 262
raccoon (*Procyon lotor*), 130,
  139, 224, 282
radio collar (for wolves),
  45–46, 47, 112, 201, 202
  mortality signal, 47
Raincoast Conservation Foundation, 145
rainforests, 16, 41, 141, 144. *See
  also* Great Bear Rainforest
  elk without wolves, habitat
    impact, 265–73
  of North Cascades, 34, 64
  Olympic Peninsula, 260–61
  protection of, 141
  rare wildlife species in, 141
  of Selkirks, 86, 112
  wolves of, 134–67, 178, 260–61
raven (*Corvus corax*), 106, 124,
  150, 153, 154–56
red alders, 19, 20
red fox (*Vulpes vulpes*), 224, 284
reed-canary grass, 280–81
Reid, Joel, 243–44, 245
rendezvous sites, 186–87,
  208–9, 253

reproduction and pup rearing, 21, 33, 52–53, 70–72,
  77–78, 120
  defense behaviors and, 251
  gestation, length of, 72
  pups, 21, 33, 52–53
  rate, 70, 185
  regurgitation and, 70, 115
  springtime births, 77
Reynerson, Marcus, 19, 274
Richmond, Dora, 261–62
river otter (*Lutra canadensis*),
  63, 130–31, 153, 154, 224,
  282
Robertson, Ray, 34, 35, 36,
  45–47, 116, 118
Rocky Mountain elk (*Cervus
  elaphus nelsoni*), 267, 273,
  274
Rocky Mountain Elk Foundation, 212
Rocky Mountain maple, 206,
  207
Rocky Mountains, 16, 34, 35
  coyotes in, 284
  elk habitat and, 273
  Northern, map of, 84
  predator and prey interac-
    tion, 273, 274–76
  wolf control in, 224
  wolf diet in, 128
  wolf habitat studies, 184
  wolf packs, home ranges, 35
  wolf prey in, 35, 126
  wolf reestablishment in, 87,
    178, 180
  wolf size and weight, 56
  wolves in, 35, 180
Rohrer, John, 36
Roosevelt, Franklin D., 262
Roosevelt, Theodore, 262
Roosevelt elk (*Cervus elaphus
  roosevelti*), 154, 262, 264,
  266, 267, 274–76
Ross Lake, 40–45
Russell, Dan, 45–47

salal berries, 141
Salish, 237
salmon, 16, 146
  Blue Mountains and, 215
  diet of bears and, 141, 148,
    153, 161–62
  diet of coyotes and, 290–93
  diet of wolves and, 36, 37,

    49, 77, 114, 117, 127, 132, 145,
    148–54, 175, 183, 292
  ecological value of, 36
  fish farms, 163
  hatcheries' dumping of car-
    casses, 36
  life of, 146–47, 150
  marine-derived nutrients
    and, 141, 146, 148–49
  parasite in, 152–53, 262–63
  runs, 34, 86, 146, 147, 163,
    167, 175, 292
  sea lice and, 163
  species of, 141, 147
  terminal fishery, 166–67
salmonberry, 139, 266, 274
Salmon River Mountains, 120,
  126, 153, 200, 208
  elk and wolves in, 208–15
Salmo pack, 84, 102–3, 106–7
Salmo-Priest Wilderness, 183, 185
Salmo River, 102
sandhill cranes, 208
San Juan Islands, 177, 253
Sawtooth Crest, 35
Sawtooth Range, 30, 32–33,
  35, 39
scat, 43, 47, 85, 103, 106, 120
  diet and, 41, 113, 128
  fresh kill and, 209
scent-marking behavior, 30,
  60, 63, 76
Scoular's willows, 206
seals, 114, 153
sea otter, 177
Selkirk Crest, 91
Selkirk Mountains, 35, 80–107
  caribou in, 85, 125
  conifers in, 104–5
  ecosystems and geography
    of, 85–87
  forest fires, 86–87
  grizzly bear in, 85, 102
  human-caused changes, 125
  inland rainforest of, 86, 112
  moose in, 85, 91–92, 153
  road closures in, 102, 104–5
  timber industry and, 86, 87,
    90, 94–95, 183
  wolf cache in, 115
  wolf DNA in, 87
  wolf packs in, 77, 84, 87–92,
    99, 102, 106–7, 112–13, 182–
    83, 185
  wolf return to, 87, 97, 180

serviceberry, 46, 207
Seven Devils Mountains, 201
Sheely, Brandon, 30, 31, 45
Shepherd, Jay, 92, 112
Sitka spruce, 19, 139, 265, 293
Skagit River, 40
skull and teeth of wolves,
    60-63, 78, 113, 120
skunk cabbage, 139
skunks, 224
Skykomish River, 290-93
smell, sense of, 60
Smith, Douglas, 127
Snake River, 153, 201, 202
Snoqualmie Pass, 34
Snoqualmie Tribe, 293
snowshoe hares, 86
social carnivores, wolves as, 15,
    41, 54, 244-45
  aggression toward transient
    animals, 76
  alpha male and female, 74
  babysitter wolf, 74
  competition for food and, 70
  dispersal of individuals, 76,
    79, 115, 116, 189-93
  dominance behavior, 251
  feeding behavior, 41
  feeding order, 117
  formation of new packs, 76
  hunting behavior, 65, 113-17
  life in a pack, 74-77
  lone wolves, 74, 76, 282
  pack as primary unit, 54, 74
  pack size, 74-75
  play, 54, 65, 72, 156, 232-33,
    234
  protection of kills and, 79
  seasonal migration, 77-78
  territoriality and, 73, 76-77
  the wolf's year, 77-78
Sol Duc River, 146
Spence, Gabe, 128
spotted owls, 141
spruce, 141
Steigemann, Volker and Iris,
    123, 132
Stellar's jays, 117
Stone, Suzanne, 242
subalpine forest, meadows, 25,
    85, 86, 93, 94, 100, 146, 190,
    267, 276, 285
swimming by wolves, 162,
    164-65
sword ferns, 274

Teanaway pack, 37, 49
territoriality of wolves, 76-77
Thieme, Al, 202, 203, 216, 229
thimbleberry, 206
Tia-o-quia-aht, 237, 244
timber industry, 86, 90
  in Blue Mountains, 202-3,
    206
  in British Columbia, 143
  caribou decline and, 94-95, 96
  clearcutting and ungulates,
    126, 182, 193, 286
  Colville National Forest as
    progressive model, 102
  crowded conifers and, 102,
    104-5
  moose and, 92, 94, 182, 185,
    194
  in Oregon and Washington,
    259, 260
  roads created by, 44, 90, 91,
    102, 103, 143, 183, 193, 206,
    265, 292
  ungulate barrens and, 154
  on Vancouver Island, 145,
    154, 158, 163, 285-86
Tofino, British Columbia, 157,
    158, 235
tracks
  black bear, 69, 226
  coyote, 66
  elk, 68
  mountain lion, 66
  mule deer, 66, 226
  wolf, 30-31, 40, 42, 64, 67,
    103, 106, 120, 142, 226
  wolverine, 68
trapping
  for fur pelts, 177, 195, 282
  of wolves, 87, 95, 177-78, 251,
    262, 293
Twisp, Washington, 37
Twisp River valley, 36

UNESCO (United Nations
    Educational, Scientific
    and Cultural Organiza-
    tion), 141
ungulates, 64, 117, 215. See also
    deer; elk; moose
  browsing pressure and,
    205-8, 219, 266, 273
  defensive strategies, 36, 121
  deforestation and decline,
    154

densities and low habitat
    biodiversity, 271, 275-76
  digestive system and, 70
  foot structure of, 64
  human impacts on land-
    scape and, 177, 182
  kill rates of, 117
  in Pacific Northwest, 93
  as prey, 39, 40, 47, 63, 64, 93,
    98, 103, 110-11, 113, 114, 117,
    120, 123, 126, 128, 154, 183,
    217, 271
  rebounding of, 183
  seasonal migration, 77
  sense of smell and, 127
  speed of, 64
  synchronized birthing
    (swamping), 121
  terrain and, 184
  wolves injured by, 63, 92, 73,
    152
United States Bureau of Land
    Management, 218
United States Endangered
    Species Act, 178, 242
United States Fish and Wild-
    life Service (USFWS), 34,
    237, 263
United States Forest Service
    (USFS), 46, 102
  caribou conservation,
    95-96
  leasing of public lands,
    218-21
  Lookout pack monitored,
    34
  wolf recovery program, 180,
    208

Vancouver Island, 134-45,
    156-67
  climate and geography,
    157-58
  deer decline, 154, 271
  Desolation Sound, 123, 132,
    159, 252, 253
  river otter, 130-31
  threats to environment, 163
  timber industry and, 126,
    145, 154, 158, 163, 285-86
  whale carcass, 132-33
  wolf attack in, 247
  wolf extirpation, 247
  wolves of, 77, 128, 132, 139,
    142, 144, 158-60, 162

Vancouver Island marmot
  (*Marmota vancouverensis*),
  285-86
Vargas Island, 246-47
vole, 114, 132

Wallowa County Wolf Defense
  Fund, 203
Wallowa Mountains, 201, 203,
  216, 221, 222-23, 226
Warnock, Dan, 241-42
Washington Department
  of Fish and Wildlife
  (WDFW), 15, 31, 34, 41,
  45-46
  elk and, 270
  wolf packs and, 34, 89, 90,
  92
  wolf recovery reviewed, 117
Washington State, 19, 133, 146,
  258
  beavers in, 129, 132
  black-tailed deer in, 125
  checkerboard landscape, 90
  last historic wolves in, 258
  livestock statistics, 218
  public opinion about wolf
    protection in, 37-38, 242
  timber industry in, 259
  wolf delisted as endangered,
  15
  wolf packs in, 34, 87-92,
  182-83
  wolf reestablishment in, 117,
  178, 242, 287
  wolf habitat in, 182, 188-89
  wolf management plans,
  195
  wolf population, 15, 178
  wolves and livestock in,
  224-25
Wenaha pack, 206
western hemlock, 41, 86, 139,
  141, 265, 273, 293
western red-cedar, 19, 41, 86,
  103, 139, 141, 158, 265, 293
western spring beauty (*Claytonia lanceolata*), 46
whale carcass, 132-33
white-tailed deer (*Odocoileus
  virginianus*), 182, 201
WildCoast Project, 247

Wilderness Awareness School,
  126-27
wildlife tracking, 16, 18-20, 22,
  29-30, 98-99, 103, 116, 150
  Hoh River, 274-75
  Methow Valley, 128
wild turkey, 35, 132
Willamette Valley, 191, 259, 270
willows, 19-20, 275, 276
wolf, common gray (*Canis
  lupus*), 55, 57
  adaptable behavior, 15, 132,
  141, 144, 244-45
  age, determining, 63, 73
  causes of death, 73, 75-76,
  125, 152-53, 173, 185, 193,
  194, 195, 222, 224-27, 261-
  63, 293
  chronology, 7-9
  delisting as endangered, 15
  die-off, 1800s, 153
  dispersal, patterns of,
  189-95
  DNA identification of, 31, 34
  as endangered, 44, 242
  extirpation of, 48, 87, 95, 98,
  144, 158, 180, 201, 224, 247,
  260-65, 272, 293
  genetic diversity, 180, 182
  history and evolution,
  54-55, 64, 65, 172-78
  life span, 73
  physical characteristics, 15,
  55-57, 60-65, 67, 78, 92, 113,
  120
  reestablishment of, 16, 56,
  87, 96, 97, 124, 144, 178,
  180, 182-89, 208, 218, 242,
  263, 286-87
wolf-dog hybrid, 31, 34, 239,
  246
wolf-human relationship,
  15-16, 18, 22, 48-49, 128,
  230-53
  attack by wolves, 239, 246,
  247-51
  communication and,
  245-47
  as competing carnivores,
  214, 215
  contemporary perspectives,
  240-44

domestic dogs and, 252
factors in, 174, 202-3
habituation and food con-
  ditioning, 247, 250-52
human contact with wolves,
  37, 39, 104-5, 120, 194, 195,
  234-35, 243-53
humans compared to
  wolves, 37, 39, 244-45
politics and, 15, 95, 174, 180,
  218, 221, 224-25, 236, 240,
  241, 242, 244
public opinion, 37-38, 175,
  202-3, 204, 229
safety tips, 248-49
traditional perspectives,
  237-40, 260
urban-rural split on wolves,
  241-42, 260
wolf as symbol, 236-40, 244,
  246
wolf population and, 183
wolf reintroductions,
  debate over, 180
wolves killed by, 125, 173,
  185, 193, 194, 195, 222, 224-
  27, 261-62, 293
Wolf's Tooth, The (Eisenberg),
  124
wolf subspecies, 180-82
  Canis lupus crassodon, 159
  Canis lupus nubilus, 180
  Canis lupus occidentalis, 180
  map, 181
wolverine (*Gulo gulo*), 62, 63,
  68, 224, 284
Wolves of North America, The
  (Young and Goldman),
  262-63
woodland caribou (*Rangifer
  tarandus caribou*), 93
woodpeckers, 222-23

yellow pine chipmunks, 46
Yellowstone Grey Wolf Resto-
  ration Project, 127
Yellowstone National Park,
  124, 128, 180, 215, 252, 274
Young, Stanley, 262-63

Zumwalt Prairie, 201, 203

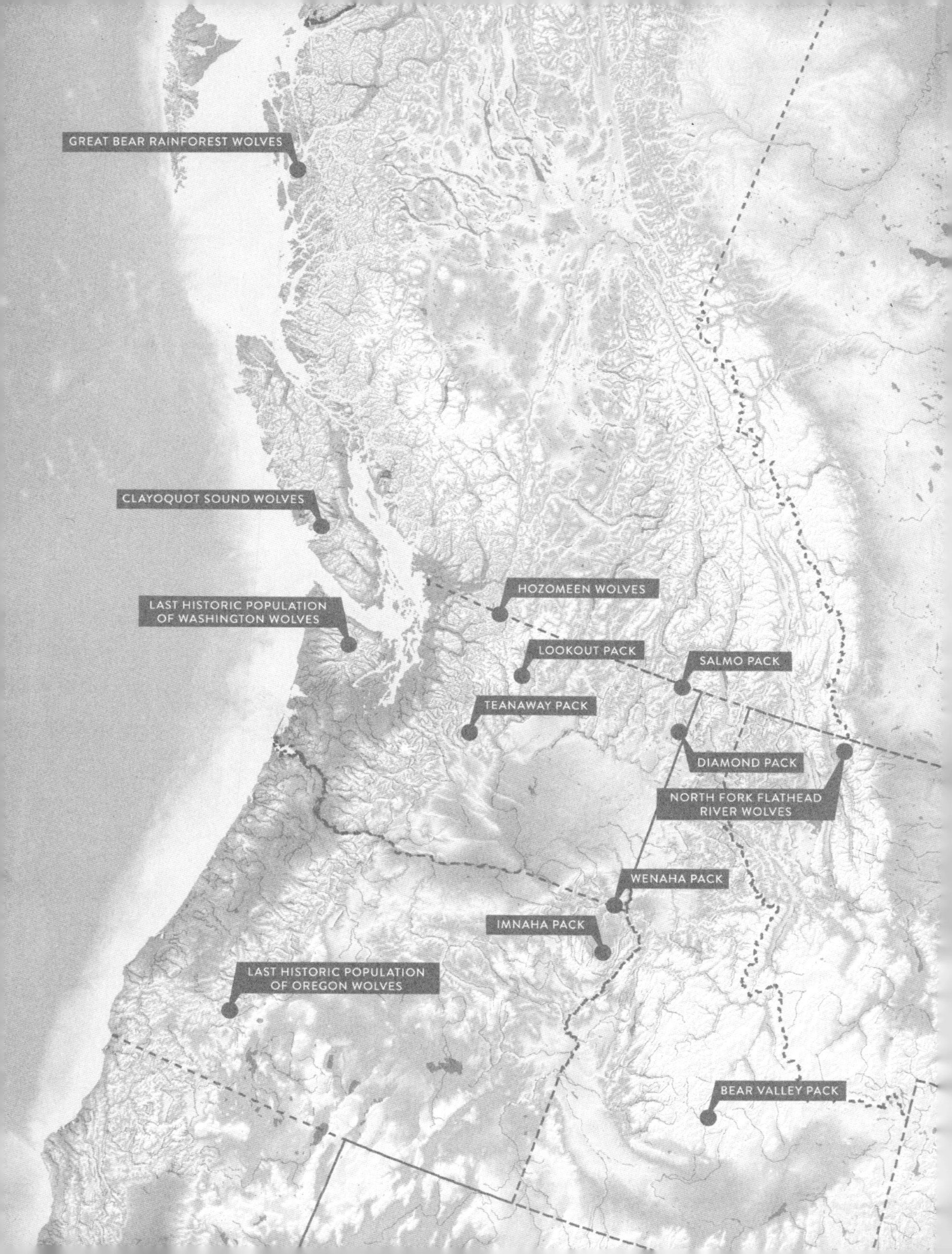

GREAT BEAR RAINFOREST WOLVES

CLAYOQUOT SOUND WOLVES

LAST HISTORIC POPULATION
OF WASHINGTON WOLVES

HOZOMEEN WOLVES

LOOKOUT PACK

SALMO PACK

TEANAWAY PACK

DIAMOND PACK

NORTH FORK FLATHEAD
RIVER WOLVES

WENAHA PACK

IMNAHA PACK

LAST HISTORIC POPULATION
OF OREGON WOLVES

BEAR VALLEY PACK